NB-IoT Use Cases and Devices

Kersten Heins

NB-IoT Use Cases and Devices

Design Guide

 Springer

Kersten Heins
Munich, Germany

ISBN 978-3-030-84975-7   ISBN 978-3-030-84973-3   (eBook)
https://doi.org/10.1007/978-3-030-84973-3

This Springer imprint is published by the registered company Springer Nature Switzerland AG
The registered company address is: Gewerbestrasse 11, 6330 Cham, Switzerland

Introduction

The "Internet of Things" (IoT) creates huge commercial expectations all over the world. Electronics industry is enthusiastic about forecasted IoT market numbers and corresponding business outlooks including a demand for billions of IoT devices. But public awareness is limited because IoT is not an appealing application or product feature itself. And IoT typically does not directly address end users. Instead, IoT is "just" an approach how to improve processes.

IoT is following the concept of "edge computing," i.e., enabling end devices to manage local data and perform local actions at a remote place while being connected to a central IoT application running on a server. This way, "things" get smart and interconnected via Internet. In fact, IoT is an umbrella term for an endless list of business opportunities to

- Maximize process **efficiency** and reduce operational cost
- Increase process output **quality**
- Improve **usability**

IoT applications are enabled by an increasing worldwide Internet coverage while data transmission cost is continuously decreasing. Thanks to mobile phones and smartphones, **cellular data networks are omnipresent and globally accepted**. Cellular networks are reliable and a solid foundation for IoT deployments. In general, this pervasive 24/7 connectivity is a key enabler for ideas how to replace inefficient processes by automated monitoring and control of remote objects—avoiding manual interaction as much as possible. And there is still much room for innovation and creativity. IoT will continue to penetrate use cases in industrial or private environments, based on networked smart devices. IoT applications will be used in production facilities, in agriculture, at home, in vehicles, etc., in an effort to improve security, productivity, failure rates, health, quality of life, service, etc.

Cellular NB-IoT technology ("NB" = narrowband) is quite radically focusing on a group of typical IoT use cases with low bandwidth requirements. As a consequence, a single NB-IoT network cell can handle 100,000+ NB-IoT devices at the same time, thus enabling massive use in high-density environments or for events.

In addition, NB-IoT devices can run from battery in unattended environments and without any maintenance for 10+ years. Thanks to outstanding power-saving mechanisms, NB-IoT enables zero-touch IoT applications which have not been feasible before. In addition, it offers wide network coverage for rural areas as well as deep network penetration into buildings even below ground level.

About this Book

This book presents the cellular wireless network standard NB-IoT (Narrow Band-Internet of Things), which addresses many key requirements of the IoT. NB-IoT is a topic that is inspiring the industry to create new business cases and associated products. The author first introduces the technology and typical IoT use cases. He then explains NB-IoT extended network coverage and outstanding power saving features which are enabling the design of IoT devices (e.g., sensors) to work everywhere and for more than 10 years, in a maintenance-free way.

The book explains to industrial users how to utilize NB-IoT features for their own IoT projects. Other system ingredients (e.g., IoT cloud services) and embedded security aspects are covered as well. The author takes an in-depth look at NB-IoT from an application engineering point of view, with focus on IoT device design. The target audience is technical-minded IoT project owners and system design engineers who are planning to develop an IoT application.

Contents

List of Figures

About the Author

Kersten Heins is a freelance technical editor and blogger (www.iot-chips.com) for IoT solutions and embedded security. After his graduation he spent 30 years in the industrial computing and semiconductor industry as a design engineer and in various marketing positions.

IoT Target Applications

IoT applications comprise smart meters and connected sheep as well as solutions for waste management or tracking of rental cars. Very often IoT means digital transformation of processes which are requiring human interaction today. Instead, connected smart will be used in order to improve user experience and increase efficiency. And they mainly interact with other machines ("M2M" = machine-to-machine), but under human supervision, of course. Processes will be automated based on consolidated IoT data, and human interaction is converted into remote sensing and remote action whenever possible.

The IoT evolution has already started, and some promising IoT target areas have been identified. Keywords are Smart Cities, Connected Driving, Asset Tracking, Agriculture, Health Care, Smart Home, Smart Grid, etc. But there are many other ideas out there, all of them contributing to billions of IoT devices forecasted to get deployed during the next years.

As a typical IoT target use case, **periodic maintenance plans** for equipment are candidates to get replaced by a smart IoT solution: instead of following a fixed schedule for control visits of a technical staff member, local sensors will track local parameters and transmit them to a central service center. This IoT data would reflect actual machine health status, so it can be analyzed remotely and trigger a tailored service action, if required. Consequently, converting this traditional use case into an IoT application will save travel cost and increase overall efficiency and reliability of maintained equipment at the same time.

In general, the "Internet of Things" consists of **connected devices** which are exchanging information with a **central IoT application** running on a dedicated server or cloud service. Collected data received from one or multiple IoT devices will be consolidated and analyzed. As a result, the central IoT application might generate and publish reports or it can take a fact-based decision and initiate remote action. This central application will also have to manage deployed IoT devices incl. their administration. On the other end, IoT devices are delivering local data originated from remote locations in the field ("edge computing"). Embedded **sensors** are used to capture relevant parameters, e.g., a room temperature or the current geo position of a tracked object. The IoT device can also take local action, either initiated by the **embedded IoT application** (firmware) or initiated by remote control of the central IoT application. For this purpose, an integrated **actuator** can be used, e.g., a power switch or motor control circuit. Figure 1 illustrates main components of an IoT application.

Each IoT application is different, but all of them require an Internet connection to a remote IoT device. Connectivity is usually based on a wireless network and a key enabler for every IoT application. But there are several wireless network technologies which can be used for IoT connectivity. In fact, different target environments might require different network technology, and a specific application scope might require network features another technology does not offer. For some of them, NB-IoT is a good choice, but not for all of them.

Independent from connectivity, each **IoT device** will have to offer application-specific capabilities to **determine, process, and transmit relevant local data**. It needs tailored sensing and control functions in combination with a certain level of "edge" computing which are not available off-the-shelf as standard products and consequently will require custom development.

Technical requirements are various, just to mention a few: some mobile IoT applications require multi-regional cell coverage with device hand-over capability from one cell to the next, e.g., for car navigation. Other IoT devices are stationary but difficult to reach, e.g., environmental sensors in rural areas or smart meters

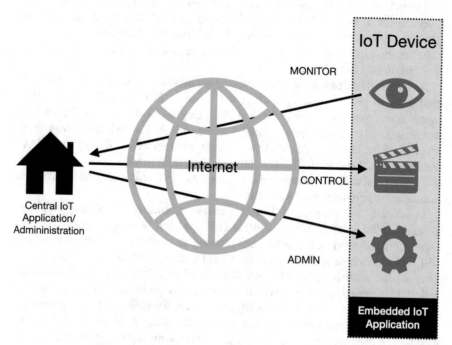

Fig. 1 IoT application and IoT device

below ground level. Some require life-long battery power. Sometimes, huge quantities of IoT devices in small areas are requiring attention, e.g., wearable IDs for visitor tracking at a big open-air event.

Wireless WAN for Internet Access

Each commercial IoT project is based on its own business conditions, and some technical parameters might be fixed. For example, if you follow an invitation to a public tender for a national smart meter project, used network technology and infrastructure will be predefined and not negotiable. But in most cases and from a technical point of view, various wireless technologies will be able to meet requirements of a specific IoT application, not just one. Figure 2 provides an overview of popular wireless technology options focusing on transmission range versus data rate. NFC with very short transmission range is used for contactless tokens rather than for IoT applications. Satellite communication offers unlimited network coverage, but too expensive for massive IoT projects. Bluetooth, WiFi, and Zigbee are no wide-area networks, but can be used locally as a secondary network in combination with an IoT gateway (see section "NB-IoT Gateway"). Category **LPWAN** (= "Low-Power Wide Area Network") is for long-range communications at comparably low data rates. Cellular technologies NB-IoT or LTE-M are LPWANs as well as competing technologies LoRaWAN and Sigfox which are no cellular networks and operate in unlicensed frequency spectrum.

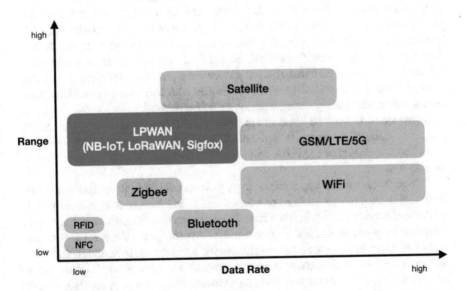

Fig. 2 Wireless connectivity technologies

But besides technical aspects, selection of an appropriate connectivity approach also requires consideration of commercial and strategic parameters. One of them is **total cost of ownership** (TCO). When you compare connectivity cost and you look at IoT service packages using cellular networks, they look expensive because usage fees apply. Others appear as free-of-charge. Background for applied usage fees is that cellular network technologies developed under 3GPP umbrella are using the so-called **licensed spectrum**. This means that network operators have acquired part of the public frequency space from national or regional authorities. Foundation of the business model for operators to offer cellular networks was driven by mobile communication, i.e., transmission of voice rather than IoT data. With more than five billion subscribers worldwide, this is an unparalleled success story and a win-win situation for operators as well as for mobile phone users.

Data transmission was part of the offered service right from the beginning: SMS (short message service) was specified back in 1986. GPRS as a packet data service embedded in cellular radio network GSM was opened in year 2000. Dedicated IoT services like (NB-IoT or LTE-M) have been specified a couple of years ago, and network deployment is ongoing (see section "From GSM to Cellular IoT" and chapter "NB-IoT Network Deployment"). In fact, cellular network operators do not have to create dedicated new NB-IoT networks from scratch. Instead, NB-IoT was designed in a way that operators can "upgrade" existing GSM and LTE networks and offer NB-IoT data services on top [1] of them. For sure, users of IoT cellular networks will be charged, e.g., per connection and/or per transmitted data volume. But, considering low investment for network infrastructure on operator side, and infrequent IoT traffic and low data volumes on IoT user side, usage fees should be reasonable and competitive.

In comparison with **unlicensed network technologies like WiFi, Sigfox, or LoRaWAN** there is one major differentiator: cellular network operators are independent service providers. They own the complete network infrastructure and will provide IoT connectivity essentially everywhere as a service. In contrast, an unlicensed connectivity infrastructure will be installed case-by-case at locations where IoT connectivity is needed for specific purpose, e.g., in office building or at home. These installations are usually owned by the user who will have to provide network access points, user management, security, sufficient network capacity, etc. Dedicated maintenance staff must be hired to ensure service delivery, software updates, support, etc. Related investments and operation expenses are impacting TCO and should be calculated for a realistic comparison of licensed vs. unlicensed connectivity solutions.

Critical in-house IoT applications (e.g., process control in industrial production facilities) do not tolerate failure or unavailability and require a high level of reliability and service quality, so they have to be designed with extra care. Very often, a process responsible prefers to **own** critical network installations for full control and in an effort to guarantee sufficient capacity, low latency, robustness, and availability. But an external operator model for parts of a critical network infrastructure might help if a stringent **service level agreement (SLA)** has been applied, e.g., for external connections or on-site wireless network coverage. In fact, SLAs are offered by cellular network operators—combined with related auditing services. An **IoT gateway** approach can be used to combine benefits of WAN and in-house LAN infrastructures for IoT applications. See section "NB-IoT Gateway."

For massive IoT deployment scenarios, **scalability** is a critical requirement. This means that used network must be able to handle a growing number of users or changing coverage conditions. For example, some countries have initiated national roll-outs of smart meters in order to track power consumption in households. For this purpose, country-wide and robust coverage network coverage is mandatory because IoT devices might be located virtually everywhere and require reliable connectivity even under worst conditions, e.g., if installed in underground basements. For this kind of large-scale IoT projects, benefits of a cellular network technology are unquestioned.

Another major benefit of a cellular IoT solution is that used network infrastructure is **always up-to-date** and future-proof. Cellular networks are based on **global** standards with very large industry support including manufacturers, network and service providers. **Long-term support and reliable service delivery** are quasi-guaranteed because of billions of users all over the world are using it.

Each IoT project is aiming at different environments and facing different conditions at target locations. One of the key prerequisites for a successful IoT concept is a clear understanding of **requirements for deployment of IoT devices**. How many will be installed and where? In general, cellular network devices are easier to deploy because connectivity depends on network coverage only: if the IoT device is within reach of a network cell, Internet connectivity will work for this device. At first sight, using an existing network free-of-charge at point of deployment looks like a reasonable approach. But will it work, will local network coverage reach all locations and provided capacity guarantee for reliable service of all IoT devices? In fact, an existing network installation can help if well-known and under control. In all other cases or in case of uncertainty resp. mission-critical IoT applications, use of cellular IoT networks should be considered—at least as a fallback option. If cellular network technology is used, no specific knowledge nor re-configuration of local IT network infrastructure (e.g., routers) will be necessary. This might be beneficial because **installation of cellular IoT devices is independent from existing local IT landscape**, if any.

From GSM to Cellular IoT

Originally, GSM (= "Global System for Mobile Communications") has been developed for digital mobile phones to be used across national borders. International cooperation work started 1982 in Europe, and first GSM networks were implemented from 1992 onwards. Since then, GSM standards have been adopted by national mobile network operators (MNOs) and deployed successfully. In 1995 around ten million users were using GSM-compliant mobile phones. Standards were evolving from GSM Phase 2 to UMTS/3G, then LTE/4G and now 5G in an effort to improve service at higher data rates and more features.

The **GSM Association** (short: "GSMA," URL: www.gsma.org) represents the interests of **mobile network operators (MNOs)**. More than 750 MNOs are full GSMA members and another 400 companies in the broader mobile ecosystem are associate members including handset and device makers, software companies, equipment providers, and Internet companies. According to GSMA we already surpassed five billion users worldwide in 2019. These are mainly mobile phone users, but IoT machine-to-machine number of connections already reached 12 billion in 2019 and supposed to reached 24.6 billion in 2025 [2].

Technical specification work moved from European ETSI to **3GPP** (Third Generation Partnership Project, URL: www.3gpp.org). 3GPP is the global standardization organization behind the evolution and maintenance of GSM, UMTS, LTE, and 5G cellular radio access technologies. 3GPP work is coordinated by regional organizations representing Europe, USA, China, Korea, Japan, and India. Since its start in 1998, 3GPP is publishing work items in release cycles, Release 16 was issued in 2020. Each release contains a set of features providing functionality across GSM, UMTS, LTE, and 5G and making sure that these technologies will coexist and interoperate.

Cellular network technologies developed under 3GPP umbrella are using the so-called **licensed spectrum**. This means that MNOs have acquired part of the public frequency space from national or regional authorities under the condition to provide a connectivity service to the public. A licensed frequency band is supposed to be available globally in order ensure worldwide presence of a service. For example, the huge success of GSM is based on the fact that GSM 900 MHz band is available in most parts of the world. This will support mobile IoT applications relying on service availability beyond national borders. As a side effect, stationary IoT installations will benefit from global compliance because large production volumes of cellular IoT devices will reduce manufacturing cost.

In parallel to improvements for consumer market of mobile phones, smartphones, etc. 3GPP has evolved their cellular technologies to incorporate requirements of machine-type communication (MTC) and to address IoT use cases. In fact, some early IoT features have been offered by cellular networks since second generation, e.g., fax or SMS. But since LTE enhancements for MTC have been included in Release 13 (2016), IoT connectivity became an integral part of cellular technology.

In the early days of GSM, a voice or data connection were "circuit switched" services (CS), i.e., all radio resources were occupied during the full duration of the connection. In 1996, efforts were started to specify "packet switched" services (PS). This kind of service was able to release resp. share resources whenever no data is being transmitted. The first PS service was called General Packet Radio Service (GPRS) which was successfully offered as a commercial service since 2000. Then, GPRS was further enhanced for higher data rates and called EDGE. **GPRS/EDGE** is still supported by most cellular networks and offers worldwide coverage for IoT devices with GPRS modems. Consequently, modems with optional GPRS support are still offered for new IoT products (see section "NB-IoT Cellular Network Modules" of chapter "Ingredients for NB-IoT Design Concepts").

Push for IoT-specific improvement of cellular networks was driven by increasing customer demand and competitive LPWAN offers of LoRaWAN and Sigfox in 2015. As a result, 3GPP introduced Narrowband Internet of Things (NB-IoT or Cat NB) technology in Release 13 following ambitious objectives on coverage, capacity, and power consumption in order to provide significant improvements over GPRS/EDGE. In parallel, LTE-M (or "Cat. M") has been specified as a second cellular LPWAN technology option with slightly different application focus on mobility and higher data rates. Like NB-IoT, LTE-M was also introduced in 3GPP Rel 13. LTE-M is a stripped-down version of LTE which is natively supported by existing LTE bandwidths. LTE-M also uses the same fundamental downlink and uplink transmission schemes as LTE. Both LTE-M and NB-IoT are LTE technologies and have many similarities, but NB-IoT with its narrow system bandwidth of 180 kHz can even work in GSM spectrum and offers flexible deployment options for network operators. LTE-M was initially limited to a bandwidth of 1.4 MHz (Cat M1) but in order to meet the requirement for more data throughput, Cat M2 was introduced allowing the optional use of 5 MHz bandwidth in Release 14.

Figure 3 provides a top-level comparison of cellular IoT network technologies which are commercially available today. Legacy GPRS/EDGE is mentioned as a

	GSM EDGE	LTE Cat M1	LTE Cat M2	LTE Cat NB1	LTE Cat NB2
3GPP Release		Release 13	Release 14	Release 13	Release 14
Downlink Peak Rate	474 kbit/s	1 Mbit/s	4 Mbit/s	26 kbit/s	127 kbit/s
Uplink Peak Rate	474 kbit/s	1 Mbit/s	6 Mbit/s	62 kbit/s	159 kbit/s
Latency	700ms - 2s	50 -100 ms		1.5 - 10 s	
Duplex Mode	Half Duplex	Full or Half Duplex	Full or Half Duplex	Half Duplex	Half Duplex
Bandwidth	200 kHz	1.4 MHz	5 MHz	180 kHz	180 kHz
Device Transmit Power	20 / 33 dBm	20 / 23 dBm	20 / 23 dBm	20 / 23 dBm	14 / 20 / 23 dBm
Modem Cost		low		very low	
Reach/Range		very good		excellent	
Capacity		up to 1 mio per sq km		up to 1 mio per sq km	
Cell Handover		yes		no	
Voice Suppport		yes		no	

Fig. 3 Overview of cellular wireless standards for IoT

reference and not recommended for new designs. But due to its exhaustive global availability, GPRS/EDGE connectivity is still relevant. In fact, GPRS/EDGE can be used as a fallback option wherever coverage of preferred network is not available or not sufficient.

Each IoT application is different, and application-specific requirements might be relevant for selection of a suitable cellular network, for example

1. Max. speed at which an IoT device can deliver data (bandwidth)
2. Max. distance between an IoT device and cell tower (range)
3. Device geo location and mobility
4. Delay until data transmission has been completed (latency)
5. Power consumption
6. Implementation cost

In comparison with NB-IoT, LTE-M is ideal for mobile devices because it handles hand-over between cell towers much like standard LTE. For example, if a vehicle moves from point A to point B crossing several different network cells, an LTE-M device would behave the same as a cellular phone and never drop the connection. An NB-IoT device, on the contrary, would have to re-establish a new connection at some point after a new network cell is reached.

Both LTE-M and NB-IoT offer outstanding cell capacity incl. ability to handle up to one million devices within a square kilometer. Both are offering sophisticated power saving mechanisms enabling battery-powered IoT devices with up to 10 years lifetime. Same applies also to network coverage which—by design—outperforms reach of GPRS/EDGE modems by up to 20 dB. But NB-IoT excels in indoor industrial environments. Since NB-IoT relies on simple waveforms, NB-IoT devices can offer better building and obstacle penetration. This provides businesses with greater range and coverage perfect for IoT projects that have sensors deployed underground or in other hard-to-reach areas with poor signal.

Besides its capability to support mobile devices, main differentiator of LTE-M vs. NB-IoT is its bandwidth of 1.4 MHz (for Cat M1) resp. 5 MHz (for Cat M2) which is opening doors for IoT applications which NB-IoT cannot address. But this extra speed leads to higher modem price, whereas NB-IoT is aiming at ultra-low-cost IoT devices. By the way, LTE-M also supports voice functionality which is needed for several IoT applications, including devices with emergency call functions such as child trackers.

NB-IoT Use Cases

As explained before, an operator-managed cellular network in licensed spectrum offers reliable and sustainable Internet connectivity for IoT applications. In particular, NB-IoT technology is a good choice for IoT applications with the following characteristics:

- Battery-powered IoT devices with ultra-long lifetime
- Infrequent transmission of small data packets with low data rate
- Relaxed latency requirements
- Large number of connected devices in same cell (high device density)
- Deep indoor or underground penetration
- Low cost

NB-IoT network capabilities perfectly match connectivity requirements of many IoT applications for industrial as well as for consumer markets. Keywords are Smart Cities, Connected Driving, Asset Tracking, Fitness Trackers, Predictive Maintenance, Smart Factory, Agriculture, Health Care, Smart Home, Smart Grid, etc. (see Fig. 4). List of target applications is long and it is growing day by day. Application areas are different, but functional similarities are leading to two main categories: **Remote Monitoring** and **Tracking/Localization**. But let us first take a look at a couple of NB-IoT projects and IoT systems. In fact, these real-world examples are revealing the variety of NB-IoT target applications and the overall potential of NB-IoT:

1. **Connected Sheep in Norway**
   Problem is that farmers gathering their animals face difficulties to find all of their animals. Network operator Telia Norway and a partner company equipped 1000 sheep with collars with NB-IoT tracking modules which are used to remotely track and monitor the animals.
2. **Shell uses IoT for pipeline monitoring**
   Used IoT solution is part of Shell's "Digital Oilfield" project in Nigeria. It provides pipeline surveillance and wellhead monitoring capabilities to remote infrastructure. Shell has managed to make savings of over $1 million.

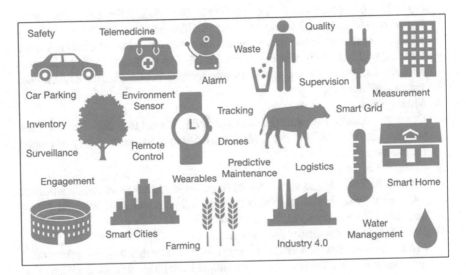

Fig. 4 Variety of NB-IoT target applications

3. **Smart parks in Las Vegas, USA**

   This initiative is driven by Las Vegas city authorities to ensure safety and to optimize city resources. For example, if somebody is in the park at night, police will be alerted so that they can take appropriate action. Utilization of park can be measured by counting visitors and find out which park assets are well accepted, e.g., a basketball court.

4. **Public space waste collection**

   In an effort to reduce waste in public space and cost of waste collection, city of Dún Laoghaire-Rathdown in Ireland decided to replace all 530 traditional litter bins and installed 420 units of BigBelly (www.bigbelly.com) IoT-enabled litter bins instead. Now, daily waste collection tours will be tailored according to transmitted fill level reports. This transformation decreased waste management efficiency by a total of 85% incl. reduction of dedicated collection staff and trucks.

5. **Protection of shipment containers**

   Company Eye-Seal (www.eye-seal.com) offers battery-powered IoT sensors monitoring the interior of a shipment container over time, i.e., access of people and temperature, humidity, etc. These data records are available online and can be used by shipping companies as evidence in case of cargo damage, theft, or customer complaints.

6. **IoT padlock**

   Company OpticalLock (www.opticallock.com) offers an IoT-enabled padlock which has a shackle that fits inside the eye rings designed for traditional locking mechanisms and provides IoT status alerts. It has sensors that detect motion, humidity, location, etc. When there is a change, the lock uses built-in cellular connectivity to send an alert.

7. **Remote Sleep Apnea Treatment**

   Up to 21% of women and 31% of men suffer from obstructive sleep apnea syndrome (OSAS). Special ventilators are an effective treatment for OSAS, but optimal air pressure level requires cloud-based computer power. Company SXT (www.sxt-telemed.it) is offering a connected OSAS ventilator for this telemedicine application which transmits relevant local data and receives control data via cellular network.

8. **Mouse Trap**

   Company TrapMe (www.trapme.dk) is offering a snap trap that reports online whether there is a catch or not. This significantly increases efficiency of related business operations.

9. **"Internet of Trees"**

   Dryad's LPWAN-based sensor network (www.dryad.net) can detect wildfires and provide valuable insights into the microclimate and growth of the forest, i.e., actionable information for forest owners.

10. **Drinking water dispenser for Africa**

    eWaterservices (www.ewater.services) is using IoT technology to build up and maintain a water infrastructure in developing countries. Installed eWater dispensers are delivering clean water on a tap-and-pay base. Revenues are used for system maintenance and to keep the water flowing. Dispensers are IoT-enabled to manage of micro-payments and water usage.

11. **Air Quality Index (AQI) Sensor**

Company libelium (www.libelium.com) is offering its Waspmote environmental IoT sensor in a robust waterproof IP65 enclosure with specific external sockets to connect the sensors, optional solar panel, etc. Several configurations are available incl. electro-chemical gas sensors providing extremely accurate ppm values and a high-end dust sensor.

12. **Smart Parking in Palo Alto, California**

City of Palo Alto (in Silicon Valley, near San Francisco) is using NB-IoT for a Smart City approach to avoid traffic congestions caused by cars go around and round the same block, in search of parking spots. Thus, sensors were installed at all the parking spots around the city. These sensors pass the occupancy status of each spot to the cloud. Any number of applications can consume that data. It can guide the drivers through the shortest route to an open spot. See https://www.paloaltonetworks.com/cyberpedia/smart-cities-in-the-age-of-5g-and-iot

13. **Visitor Engagement Tracking**

Wearables can be used to track user movement and behavior. Company Sirqui (https://corp.sirqul.com) uses IoT techniques to understand visitor preferences, walking paths, duration of stay, queue lines of big events in stadiums or arenas. These insights can be used to take appropriate action for improvement of visitor experience.

14. **Perfect irrigation for "happy nuts"**

South African nut farmer turned to LPWAN solutions by u-blox (www.u-blox.com) to manage R&K Estates Macadamia nut farm and improve efficiency and yield. In fact, periods of drought are causing "stress" to Macadamia trees. Moisture sensors in combination with timely controlled water sprayers solve the problem.

In general, for NB-IoT all advantages of an operator-based cellular network apply, e.g., reliability, availability, security. Figure 4 illustrates major potential NB-IoT target applications areas. In particular, NB-IoT key benefits in comparison with LTE-M are (1) lower cost and (2) better network coverage, i.e., wider range, and improved obstacle resistance. These differentiators are leading to a list of focus target applications which are leveraging NB-IoT particular competitive strengths. In fact, NB-IoT devices are tailored to address applications requiring

- **Battery power supply** and long lifetime (i.e., if no local power outlet is available)
- **Zero-touch operation** (i.e., no device maintenance or battery recharge resp. replacement is planned)
- **Stationary** operation, i.e., a fixed device location (or moved within same cell only)
- **Wide coverage** range, e.g., in a rural environment
- **Deep penetration** of environments with poor signaling conditions, e.g., in buildings or forests or locations below ground level
- Lower cost as LTE-M.

In general, NB-IoT target applications are latency-tolerant and focused on infrequent transmission of small data packages. Focus is on uplink traffic rather than download traffic, bandwidth and transmission delay requirements are low. IoT applications are sometimes categorized by target industries or target markets like Smart Cities, Smart Farming, Smart Home. But these categories do not describe functions to be implemented or any technical requirements. Each IoT application is different and relies on application-specific IoT frontends with dedicated sensing and control functions plus associated edge computing power and memory. Consequently, the following list is focusing on **IoT device functionality**. NB-IoT connectivity is a good choice for applications aiming at

1. **Remote Monitoring** of remote equipment, an area or a site or any kind of object in an effort for information purposes or to identify problems requiring action **or Detection** of remote objects (incl. people, animals) and determination of further details of detected objects, e.g., their identity or headcount.

   (a) **Measure local parameters** (e.g., moisture, temperature, position, acceleration, force, fill level, presence, motion, gas or liquid flow, light, power consumption, intrusion, carbon monoxide, identity, heart rate).
   (b) **Extract, determine, and send relevant, actionable data.**
   (c) **Remote Control** of local actions (e.g., in addition to monitoring/detection function of IoT device), via actuators like a motor, power switch, etc.

   - Application examples: surveillance, emission warning, asset utilization, service alert, home automation, fault detection, POS refill, energy management, tamper detection, livestock monitoring, quality management, access control, recognition, inspection of infrastructure, leakage and flood detection, occupancy monitoring, outage detection, dead reckoning, trash management, smart parking, traffic control, predictive maintenance, patient monitoring, smart lighting, performance monitoring, crop yield improvement, tank monitoring, etc.

2. **Tracking/Localization** of movable objects, e.g., assets, goods, material, inventory, containers, visitors

   (a) Locally (e.g., within a production site)

   - Application examples: logistics, supply chain visibility, shipment handling, material management**or**

   (b) Anywhere, in combination with precise GPS localization

   - Application examples: investigation of incidents of stolen or lost objects, e.g., valuable goods, shipment containers, expensive equipment.**or**

   (c) In capacity-intensive environments, i.e., large number of IoT users covered by same network cell

   - Application examples: event management, visitor behavior tracking.

In sum, we have identified **two main IoT application areas which are predestinated for NB-IoT** cellular connectivity: (1) **Remote Monitoring/Detection** with or without Remote Control, (2) **Remote Tracking/Localization**. Besides this functional difference of both application areas, Monitoring/Detection applications are characterized by the fact that each IoT device is aiming at a specific location where it has been installed (stationary operation), i.e., they are **fixed** objects. On the other hand, Tracking/Localization IoT devices are integrated into **movable** objects, i.e., objects which are not fixed to a single location.

Object Tracking/Localization

For this group of NB-IoT target applications the main objective is to **determine the current position of a movable** object. Technical requirements are different from mobile IoT applications which are following the position of a mobile object in real-time. Mobile IoT applications create more data, do not tolerate high latency, and require network hand-over of IoT devices crossing cell borders. For this kind of scenario, LTE-M is far better choice. But NB-IoT has been built to transmit small data packages case-by-case, and occasional positioning of a movable object is asking exactly for this. In fact, a movable object does not move all the time, but it might change its location—by purpose or by accident. Depending on the application environment, detecting actual position data will be done in a different way, e.g., read a coded identifier on a routed path or from limited set of possible locations the object cannot leave. In other cases, e.g., for a shipment container or a stolen piece of equipment, the actual position will be completely unpredictable and must be determined from scratch. Ultimate localization precision (i.e., geo coordinates of object position) is offered by **satellite positioning systems** like GPS or GNSS. Many manufacturers of IoT cellular network modems also provide dedicated positioning modules and combinations of both ("bundles")—as a hardware option plus related software support. See section "NB-IoT Cellular Network Modules" of chapter "Ingredients for NB-IoT Design Concepts" for further details about component selection aspects.

In general, and by nature of a moveable object, an IoT device providing tracking/localization data will have to be powered by a battery. If used in industrial environments or for logistics, batteries for tracked objects will be recharged in a timely manner and will not cause any operational bottlenecks. Otherwise, i.e., if batteries are no candidates for recharge or replacement ("zero-touch" operation), battery lifetime can be estimated based on scheduled activity periods, see section "NB-IoT Device Battery Lifetime Calculator" of chapter "Designing an NB-IoT Device" and select reasonable battery capacity according to application requirements.

Other good reason for IoT object localization are use cases where objects are moved or relocated by unauthorized people or unintentionally, i.e., they were lost or have been stolen. IoT technology in combination with a GPS/GNSS satellite positioning system can be used to identify the current position of the object and to transmit a corresponding notification message containing object geo coordinates. A suitable add-on IoT device should be prepared to protect any kind of object and for zero-touch operation during the complete lifetime of the protected object, i.e., for many years. This is why it should be independently powered by an integrated battery. For this use case (localization of a lost object), it makes sense to request actual position data only if needed (**"pull" mode**) instead of pushing it periodically by the IoT device. This will reduce redundant data transmission and save significant battery power.

For security reasons, the localizing IoT device or subsystem should be integrated/embedded into the object in a non-removable and tamper-resistant manner, e.g., during production of the protected object. Initialization of the IoT device will be required in order to prepare for field operation, e.g., for personalization purposes, configure operation mode, preselect networks, etc. Then, after deployment, the device will be in sleep mode by default and wait for external wake-up paging event from NB-IoT network. A design concept for this kind of typical NB-IoT target application (working in "pull" mode) will be introduced in section "Design Concept #2: Object Localizer" of chapter "Designing an NB-IoT Device."

Remote Monitoring/Detection

This is probably the most promising group of NB-IoT target applications. It is covering a large variety of use cases across many industries and application areas with unlimited creative potential and business perspective, even for consumer markets. Low modem cost is a key benefit of NB-IoT, see section "NB-IoT Technology" of chapter "Cellular IoT Technology." Device cost of a suitable IoT device for Monitoring/Detection use cases can be reduced even further with ultralong battery power because installed battery will often determine lifetime of the product itself, i.e., if **zero-touch operation** is planned. Battery capacity should be tailored to business requirements, e.g., for a total product lifetime of 10 years. Section "NB-IoT Device Battery Lifetime Calculator" of chapter "Designing an NB-IoT Device" is providing a calculation spreadsheet which can be used to estimate required battery capacity of an NB-IoT device.

Typical monitoring IoT applications are keeping users aware of local conditions **periodically**, e.g., once a day or every few hours, in a "push" operating mode. Alternatively, NB-IoT devices can also alert online users immediately whenever a certain threshold value has been exceeded. Similar **alert function** is also good for **surveillance** equipment which is triggering data transmission only if presence of an object or a person has been detected.

In general, NB-IoT is good for IoT devices which **remain in low-power state ("sleep") most of the time** and wake up only when scheduled or triggered by a local event. This kind of operation allows to leverage NB-IoT outstanding power saving modes. After a scheduled modem wake-up (or a local trigger event), the IoT node will reactivate other device components, perform sensing process, retrieve corresponding data, perform some edge computing tasks, and finally transmit updated

Fig. 5 Simplified activity loop for typical NB-IoT device in "push" mode

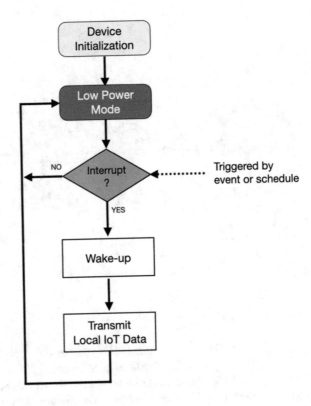

IoT data to a pre-configured external recipient. Then, the device will return to sleep as illustrated in Fig. 5.

A design concept for this typical NB-IoT target application (i.e., "push" IoT device with ultralong battery life) will be introduced later in section "Design Concept #1: Environmental Sensor" of chapter "Designing an NB-IoT Device."

NB-IoT Gateway

For some business-critical connected applications, e.g., for production process automation, customers are requiring an extra level of reliability and security, so they do not feel comfortable to handle in-house machine-to-machine (M2M) process communication between involved equipment through a shared public network. A service level agreement (SLA) offered by cellular network operators is supposed to address this kind of customer concerns, but will not be able to cover all related risks and liabilities to full extent. Consequently, customers will prefer to set up a tailored in-house local-area network (LAN) infrastructure for the critical part of the in-house M2M infrastructure.

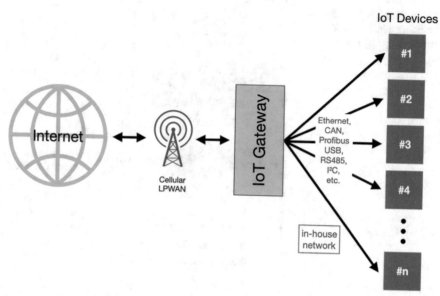

Fig. 6 IoT gateway approach

On the other hand, wide-area (WAN) connectivity might be needed to interconnect distributed production facilities or to a central cloud service. This can be done by a cellular LPWAN access point combined with a gateway function which allows to interconnect LPWAN and in-house proprietary network in a customer-defined manner. From an IoT perspective, this **IoT Gateway** might act as an IoT endpoint itself or pass messages to/from IoT devices connected via LAN or CAN bus or other interfaces used in industrial or consumer environments. Figure 6 illustrates how this would work.

Mission-critical applications in home or office environments with existing LAN connectivity are another target area for IoT gateways. For example, nation-wide roll-out of smart power meters in Germany is specifying a central device called "Smart Meter Gateway (SMGW)" to be installed in more than 40 million German households and other electricity consumers. For this application scenario, every commercial SMGW device will provide cellular connectivity as a fallback option in case other connectivity options are not available or an existing network installation fails to provide sufficient connectivity (e.g., for power meters located in underground basement). In fact, the SMGW is a quite complex IoT device providing cellular network connectivity and a couple of interfaces for internal connection of electricity meters as well as solar panels, car chargers, etc. SMGW devices have to comply with a specification which has been prepared by BSI institute (www.bsi.de) on behalf of the German government. For each commercial SMGW product an evaluation and official approval is mandatory, which is covering quality of implemented mechanism for data security and tamper protection (see also section "Security Hardware and Certifications" of chapter "Ingredients for NB-IoT Design Concepts").

Cellular IoT Technology

Cellular Network Basics

A cellular network is made up of a large number of signal areas called cells. These cells join or overlap each other. Within each cell you will find a base station or **cell tower** which sends and receives data. A mobile phone or IoT device usually connects to the closest available **base station**, if not congested by other user devices. Base stations are gateways for wireless data to the MNO network. For IoT devices, a connected base station is acting as the entry point for Internet communication (Fig. 1).

Cells are owned and managed by MNOs which are usually private companies. Acquired license provides exclusive usage rights for a certain carrier frequency range to the MNO. Connectivity services offered by an MNO will be usage-based, i.e., users will have to agree on commercial conditions and sign a corresponding contract with selected MNO, i.e., a **subscription plan** applies. Purchase of a national cellular license might include some governmental obligations, but normally each MNOs will take **network coverage** decisions based on business perspective, i.e., based on customer demand. With other words: an MNO will not extend network coverage into areas where nobody will use it. In addition, each cell has a limited **capacity**, i.e., it provides only a limited number of **channels** for simultaneous user connections. As a result, each MNO will create a cell floorplan and manage cell deployment accordingly—in an effort to offer good network coverage and service for all subscribers.

A cell tower is a platform where antennas and other hardware are being mounted. Physically, this could be a dedicated mast or a building. In general, cell range (or cell size) depends on used antenna and applied output power (which might be restricted/limited by regional law). In rural areas, a single network cell will be able to cover an area of up to 35 km, but density of cells will be low. With help of a range extender, Australian MNO Telstra even claims to be the "first operator globally to implement a 120 km radius on NB-IoT" [3]. By nature, cell coverage area will be small in urban areas because high-frequency radio waves cannot easily pass

K. Heins, *NB-IoT Use Cases and Devices*, https://doi.org/10.1007/978-3-030-84973-3_2

Fig. 1 Network cell to
Internet

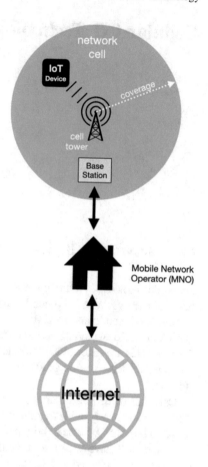

building walls easily. On the other hand, high population in urban areas means a lot
of users might request service at the same time. This means that a new connection
request would be rejected by a cell if its capacity is exhausted. This gap will be
compensated by additional cells within reach of a user terminal, i.e., it might not
connect to the closest cell. Instead, it will allocate the next available channel offered
by another cell within reach. This way, a high density of small cells in urban areas
will enable massive connectivity (Fig. 2).

In general, device **power consumption** for uplink data transmission correlates
with distance to cell tower. Closer distance resp. a small cell means that less output
power is required by a user device to send data. At least for battery-powered stationary IoT devices, location resp. distance to attached cell tower might have a significant impact on battery lifetime. In urban areas, so-called femtocells (10 m range) or
picocells (200 m range) are used to cover buildings which are also supporting IoT
deployments.

Fig. 2 Cellular
network grid

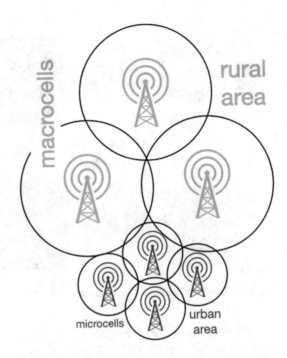

Modem Interface

Before we start to look closer at cellular network technology, we have to travel back
in time for a minute and recall the original meaning of some telecommunications
terms and mechanisms. For example, in technical documentations of NB-IoT cellular networks modules still terms like DTE (Data Terminal Equipment) or MT
(Mobile Terminal) are being used although not quite applicable to M2M applications of today. Another example are so-called AT commands which been invented
back in 1981 to control Hayes modems to transmit data via phone line at a data rate
of 300 baud. At that time, AT control commands have been used by a computer
terminal for dialing and hanging up. Modem connection was a 9-pin RS-232 serial
interface using ±12 V levels which is not common practice any more, but UARTs
and RS-232 signals (TxD, RxD, CTS, RTS) are still being used by modern IoT cellular network modules. Figure 3 is explaining both use scenarios and some common
terms and acronyms.

Originally, data transfers always have been initiated and controlled by a human
sitting in front of a computer or terminal, and user information was entered or prepared for transmission. This communication scenario still exists, esp. with mobile
terminals, i.e., smart phones. This is why an end device is sometimes called MT
(mobile terminal) or DTE, even if—in case of machine-to-machine communication
(M2M)—an IoT application takes over and "replaces" the human user. In fact, the
IoT application is an embedded software program running inside the IoT device

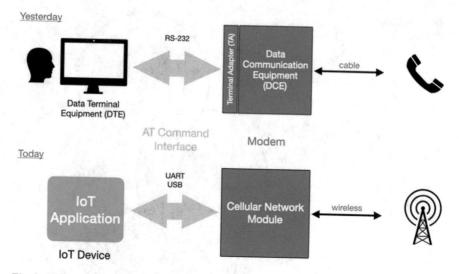

Fig. 3 AT Command Interface—yesterday and today

(see chapter "Ingredients for NB-IoT Design Concepts"). Term "modem" is still valid because this component is still used to convert digital data for telecommunication, but transmission medium telephone line has been replaced by wireless cellular network. Term "Terminal Adapter (TA)" was used for the connection part of the dial-up modem in the past, and still sometimes used as a misleading name for the **AT command interface** function of a cellular network modem.

AT commands are still used as a standardized control API for cellular network interface subsystems, which are key components for IoT devices incl. NB-IoT (see section "NB-IoT Cellular Network Modules" of chapter "Ingredients for NB-IoT Design Concepts"). AT-based communication is using UART or USB serial interfaces inside the IoT device, mastered by a dedicated application MCU.

Standard AT commands are used to control modem functions and network services. They are specified by 3GPP TS 27.007 "AT command set for User Equipment (UE)." Initial release was published back in 1999. Further evolution of the 3GPP standard and additional AT commands is reflected by newer releases and further specifications. Consequently, TS 27.007 is being updated regularly. From an application point of view, all standardized AT commands should be available and corresponding functions should have been implemented by every commercial cellular modem. In addition, related network services should have been implemented by the network infrastructure as well, otherwise functions requiring cooperation with network (esp. NB-IoT power saving mechanisms) will not work as expected.

On top of standardized functions, manufacturers of IoT cellular network modules can add extra AT commands to control proprietary functions or to utilize complementary software functions. In fact, a **cellular network module** today is a highly integrated, mixed-signal MCU-based subsystem which is—besides other functional elements—also including the modem function. Operating software is field-updatable and sometimes even customizable, i.e., open for application software. Consequently, most manufacturers are offering cellular modem extensions like Internet software stacks and communication protocols. In practice, the IoT device can obtain a virtual IP socket from the network module which is acting as a TCP/IP endpoint on behalf of the device. Or, for example, secure end-to-end communication can be handled by the module via embedded SSL/TLS stack.

From an IoT application point of view, this is added value because customer development can focus on application expertise rather than standard connectivity ingredients. This kind of software building blocks are bundled with most IoT cellular network modules, and they are supported by a proprietary set of AT commands. Also, IoT protocols like MQTT or FTP function for data transfer or firmware update are not standard and handled differently by manufacturers.

For NB-IoT device designers, selecting a powerful network module with a comprehensive AT command set is a key enabler for fast turnaround and time-to-market.

During field operation, mentioned embedded IoT application will submit AT commands to the modem via UART system call. But during development or for evaluation or test purposes, users can also issue AT commands via a hyper terminal using a command-line interface, e.g., on a Windows or Mac PC.

Examples

1. Check if device is connected to a network and registered:
Command:
`AT+COPS?`
Response:
`+COPS: 0,0,"vodafone IT",7`
`OK`

2. Change operating LTE band (for next registration attempts):
`AT+UBANDSEL=1800,2100,2600`
Response:
`OK`

A dedicated section "AT Command Interface" of chapter "Ingredients for NB-IoT Design Concepts" will take a comprehensive look at AT command syntax and provide more examples.

NB-IoT Technology

Until early 2015, GSM/GPRS/EDGE had been the main cellular technology of choice for serving wide-area IoT use cases. At this time, GPRS was a mature technology with low modem cost. But market demand for low-power wide-area networks (LPWANs) was increasing and GPRS was challenged by alternate technologies in unlicensed spectrum like Sigfox, LoRa. Anticipating this new competition, 3GPP started a feasibility on "Cellular system support for ultra-low complexity and low throughput Internet of Things" [4].

These efforts resulted in ambitious objectives for cellular IoT features in a post-GPRS era. Main goals were as follows (Fig. 4):

1. Improve cell **coverage** for rural areas and **penetration** of buildings down to ground level.
2. Reduce **device complexity and cost** in order to enable massive IoT applications.
3. Network **latency** should not exceed 10 s.
4. Achieve high **capacity** for a massive number of IoT devices generating a small amount of data.
5. Minimize device **power consumption** to achieve 10 years lifetime of a 5 Ah battery.
6. **Reuse existing LTE network** infrastructure and upgrade with new IoT features by software only.

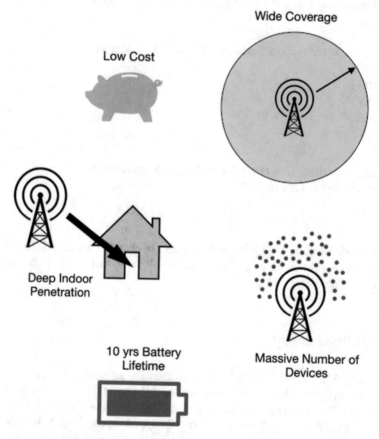

Fig. 4 NB-IoT objectives

As a result of related standardization work and worldwide agreement, **NB-IoT** (= "Narrowband IoT") has been defined as an add-on to 4G/LTE in 3GPP Release 13.

Network Deployment Options

For network operators, NB-IoT has been designed with various deployment options in mind. In a stand-alone scenario, an NB-IoT carrier would replace a GSM carrier. With a carrier bandwidth of just 200 kHz (same as for GPRS), an NB-IoT carrier can also be deployed within an LTE carrier or in an LTE guard-band. Available options are illustrated in Fig. 5.

Stand-alone deployment in a GSM band is an option for operators running GSM and LTE networks in parallel. In this case, one or more GSM carriers can be "refarmed" to the LTE network so that they can be used to carry NB-IoT traffic.

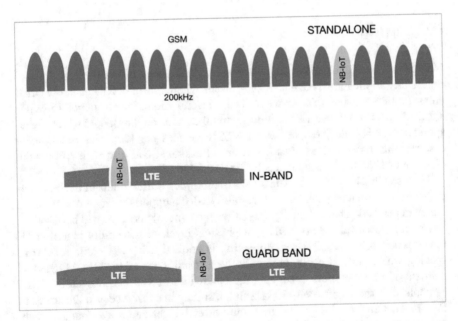

Fig. 5 Spectrum usage deployment options

Later, when time is right, the MNO can refarm his complete GSM spectrum for use by LTE. According to Ericsson [1], a leading manufacturer of cellular network equipment, this migration is a seamless process even with deployed NB-IoT carriers in the GSM spectrum as it will not impact NB-IoT devices at all and will continue to work in the LTE spectrum afterwards.

In general, for LTE operators the **in-band** option probably is best choice because it allows flexible assignment between LTE and NB-IoT. So, for example, an operator can start with a stand-alone deployment of its NB-IoT service in a GSM band and subsequently migrate these NB-IoT bands to an in-band LTE deployment depending on increasing customer demand.

A third alternative is to focus deployment of NB-IoT in LTE **guard-bands** which are adjacent to each LTE carrier. In fact, physical NB-IoT layers have been designed to coexist in LTE guard-bands without causing interference. Guard-band deployment can be done without affecting the capacity of the LTE carrier.

All three deployment scenarios are transparent to non-NB-IoT devices. Consequently, LTE devices that do not implement NB-IoT functionality simply do not see the NB-IoT channel inside the main LTE bandwidth or in the guard-band. At the same time, legacy GSM devices will not see an NB-IoT carrier if used alongside 180 kHz GSM carriers. Such devices will only see noise where NB-IoT is active [5].

By the way, with 3GPP Rel. 15 the coexistence of 5G with NB-IoT has been secured. It provides the opportunity to introduce 5G in carriers where NB-IoT is already in operation. In future, existing NB-IoT deployments can continue to provide IoT connectivity in 5G networks as a low-end and low-complexity option [6].

Cell Capacity

NB-IoT has been designed to support massive deployment of IoT devices, e.g., for Smart City applications, but also for big events where lots of people gather in small areas (with wearable IoT devices). The capacity requirements target in 3GPP TR43820 [4] for Release 13 has been set to 40 devices per household which corresponds to 52,500 devices per cell or 60,680 devices per km^2. This requirement assumes that base stations are located on a hexagonal cell grid with an intersite distance of 1732 m resulting in a cell size of $0.87m^2$ (Fig. 6).

How can this be achieved? The idea is to multiplex traffic of lots of devices altogether in an efficient manner. As a result, NB-IoT supports a range of data rates which depends on channel quality (signal-to-noise-ratio) and allocated bandwidth.

For the **downlink** (short: "DL"), at physical layer, NB-IoT fully inherits LTE transmission scheme. QFDM multiplexing is applied using a 15 kHz subcarrier spacing with a normal cyclic prefix CP. As a result, each of the QFDM symbols consists of 12 subcarriers occupying the bandwidth of 180 kHz. Seven QFDMA symbols are bundled into one 180 kHz slot, resulting in a resource grid illustrated in Fig. 7. All CIoT devices share the same power budget and may simultaneously receive base station transmissions.

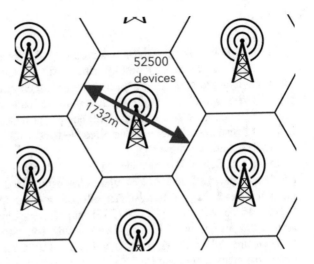

Fig. 6 Cell capacity of 60,680 devices per km²

52500 devices

1732m

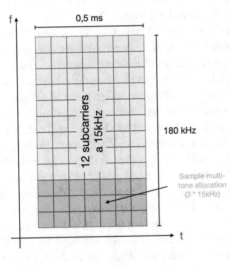

0,5 ms

12 subcarriers a 15kHz

180 kHz

Sample multi-tone allocation (3 * 15kHz)

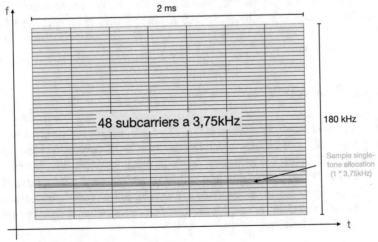

2 ms

48 subcarriers a 3,75kHz

180 kHz

Sample single-tone allocation (1 * 3,75kHz)

Fig. 7 Resource grids for 15 and 3.75 kHz subspacing

In the **uplink** (short: "UL"), however, each device has its own power budget which can be combined by multiplexing traffic of several devices in same cell. These devices can concatenate their transmission energy to a narrower bandwidth [1]. The idea is to allocate only small amounts of bandwidth to specific devices in order to increase overall capacity without performance degradation. Practically, instead of resource blocks with an effective bandwidth of 180 kHz, NB-IoT allocates subcarriers (or "tones").

In the UL, SC-FDMA mechanism is applied, either with a 3.75 kHz or 15 kHz subcarrier spacing. The 3.75 kHz option has been added exclusively for NB-IoT, it increases **flexibility of bandwidth allocation** even further. The LTE eNodeB decides which one to use [7]. For 15 kHz subcarrier spacing the UL resource grid is identical as for DL, but for 3.75 kHz it has a different structure:

There are still 7 OFDM symbols within a slot. But, according to the OFDM principles, the symbol duration for 3.75 kHz subcarrier spacing has four times the duration compared to 15 kHz, which results in a slot length of 2 ms.

Each NB-IoT device can be scheduled on one subcarrier ("single-tone") or more subcarriers ("multi-tone") in the uplink. Smallest amount of bandwidth to allocate to a device is a subcarrier of 3.75 kHz. Allocation of multiple subcarriers works only with 15 kHz ($n \times 15$ kHz, $n = [3, 6, 12]$) and will increase transmission data rates of selected devices accordingly. This option is adding extra flexibility in cellular environments handling devices both in good and in bad coverage areas. Devices enjoying good coverage and configured for multi-tone will be able to finish data transfer 12 times more quickly (vs. single-tone) and release their allocation to others in a shorter period of time. This approach will add capacity in certain macrocell scenarios, but will not be required in areas with a dense cell grid or other good coverage scenarios, e.g., where all devices are located within the standard LTE cell range.

According to simulations performed by network equipment manufacturer Ericsson [1], evaluations have shown that a "standard deployment" can support a **density of 200,000 NB-IoT devices within a cell**, i.e., achieved capacity is exceeding original goal by far (four times higher). Observations in real-world mass deployments will provide further insights, but for the time being it seems that NB-IoT multiplexing scheme in combination with adjustable data rates will meet capacity goals.

Coverage Extension

Another important NB-IoT objective is to achieve *adequate* coverage at a maximum coupling loss (MCL) of 164 dB which represents a **20 dB coverage enhancement** compared to GSM/GPRS. Practically, this will result in wider data transmission range as well as deeper penetration into buildings, underground assets, tunnels, etc. In fact, 20 dB improvement corresponds to a **sevenfold increase in coverage area for an open environment**, or roughly the loss that occurs when a signal penetrates the **outer wall of a building**.

The reduction of NB-IoT bandwidth already helps a lot because the LTE eNodeB maintains the same transmission power as in LTE case (43 dBm). In fact, it increases the power density ratio (PSD). In uplink, NB-IoT bandwidth can go down to 3.75 kHz, GPRS is at 200 kHz. Consequently, PSD ratio is 5.3 which is around 7 dB which is a first contribution to coverage enhancement goal of 20 dB.

Second contribution is **repeated transmission of data packets** depending on individual radio conditions, i.e., NB-IoT devices in difficult areas will retransmit IoT message several times in order to increase eNodeB chance to decode the received message correctly. For this purpose, three **coverage enhancement (CE) levels** have been specified: CE0, CE1, and CE2. CE level 0 represents the standard LTE coverage level, and CE level 2 corresponds to the worst case, where the coverage may be assumed to be very poor.

Note: It is up to the network how many CE levels are defined. A list of power thresholds for the received reference signals is broadcasted in the cell for each CE level.

The main impact of the different CE levels is that the messages have to be repeated several times. In case of CE0, since we are in standard LTE conditions, the device will not have to repeat the data. In CE1 the device would be out of LTE coverage, then it needs to repeat the packets 2, 4, 8, or 16 times; finally, in CE2, we have the worst channel conditions, so the device must repeat data 16, 32, or 128 times (Fig. 8).

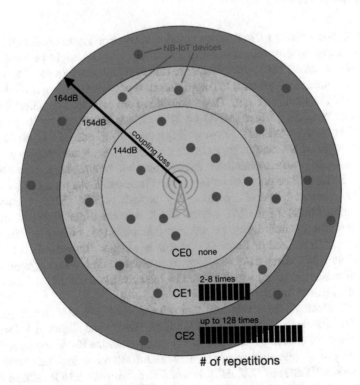

Fig. 8 Coverage enhancement levels (CE)

The repetition mechanism used by NB-IoT is a **hybrid automatic retransmission request (HARQ)** scheme. HARQ is a combination of error detection (ED), forward error correction (FEC), and automatic repeat request (ARQ) error-control. ED and/or FEC information will be added to each IoT message prior to transmission in order to detect and correct an expected subset of all errors that may occur, while the ARQ method will lead to retransmissions as a fallback in case of uncorrectable errors in original message. In HARQ type 1, an erroneous data frame will be rejected and will be re-transmitted until received error-free or fully corrected or when the maximum number of retransmissions is reached. For NB-IoT uplink transmission, a more sophisticated version is being used: HARQ type 2. In fact, HARQ 2 is **alternating embedded ED and FEC information** in re-transmitted data frame case-by-case. FEC bits are only transmitted on subsequent retransmissions as needed, i.e., re-transmitted will be shorter for devices facing good signal conditions. In this case, missing FEC info will reduce cell capacity required for coverage enhancement.

A repetition factor of two results in less than 3 dB. Thus 128 repetitions are around 13 dB [8]. So, adding these 13 dB to mentioned 7 dB gain due to increased PSD ratio, NB-IoT achieves target of 20 dB coverage enhancement vs. GSM/GPRS.

Network Selection

By nature, operation of a CIoT device requires connection to a network cell. At least in urban areas, not just one but multiple cellular networks might be available, so which one to select? Selection criteria are different and closely related to network coverage requirements of each IoT application, esp. where devices will be located and if they are mobile or not. These considerations will lead to selection of an appropriate network partner (MNO or MVNO, see also section "Roaming and MVNO" of chapter "NB-IoT Network Deployment") to ensure that **all** IoT locations are supplied with sufficient network coverage (i.e., signal strength and quality). Consequently, an associated subscription plan will be already in place whenever an CIoT device will be powered on for the first time and looks for a cell to connect to. In fact, selected network partner will provide a SIM (SIM = Subscriber Identification Module) for CIoT device which is inserted during production and usually remains inserted during complete product life (see further information in extra section "SIM Card or Embedded eSIM" of chapter "Ingredients for NB-IoT Design Concepts"). Each SIM also contains a unique number IMSI (international mobile subscriber identity) which is used by the network to identify the user of the IoT device, i.e., the owner.

For cell selection, the device will first refer to several PLMN entries stored in SIM. PLMN stands for ("Public Land Mobile Network") and allows to differentiate cellular networks offered by different MNOs using a unique PLMN code. A PLMN code is not just for operators *owning* a network infrastructure incl. cell towers, etc. like Swisscom (PLMN is "228 01"). Instead, also "virtual" MNOs (aka MVNOs) have their own PLMN identifier. So, for example, the PLMN code for specialized IoT-MVNO company Emnify is "295 09."

Usually, after power up or after relocation from an uncovered area, a device will try to **connect again to the saved PLMN** it was previously registered on. If this PLMN is not available or connection attempt fails, another PLMN can be selected, either automatically or manually, depending on the PLMN priority information stored in SIM and/or signal strength. If automatic network selection mode is disabled, the user (resp. IoT application) can select from a list of choices presented by the modem.

In general, many aspects of cell scanning and registration procedure are user configurable and can be tuned via user API offered by cellular network modules (see section "Modem Interface") according to application needs. A stationary NB-IoT device, for example, might boot only once (during installation after deployment) and will remain camped on a specific cell during its complete product lifetime. In this case, also roaming can be disabled by default.

In order to support selection process, essential network system information (SI) is broadcasted by each LTE eNodeB to all user devices within reach. It contains a **Master Information Block (MIB)** and a couple of **System Information Blocks (SIBs)**, here is a short description of first blocks to be considered:

System Information Block	Content
MIB-NB	Operation mode info (stand-alone, in-band, guard-band), etc., information about SIB1-NB scheduling
SIB1-NB	Cell-access related information such as **PLMN, tracking area and cell identities**, access barring status, **thresholds for evaluating cell suitability**, etc., scheduling information regarding other SIBs
SIB2-NB	**Radio resource configuration (RRC) information** (incl. RACH-related configuration) for all physical channels that is common for all devices
SIB3-NB	Parameters required for intra-frequency, inter-frequency and I-RAT cell reselections
SIB4-NB	Information regarding INTRA-frequency neighboring cells (E-UTRA)
SIB5-NB	Information regarding INTER-frequency neighboring cells (E-UTRA)

In general, a user device should store a valid version of MIB-NB, SIB1-NB, and SIB2-NB through SIB5-NB. The other ones have to be valid if their functionality is required for operation. For instance, if access barring is indicated in MIB-NB, the user device needs to have a valid SIB14-NB.

Access Barring (AB) is an access control mechanism adopted in NB-IoT. It allows a PLMN to restrict, allow, or delay access according to assigned access class of a device (which is stored in SIM). For LTE, 15 access classes have been specified, 10 "normal" classes and 5 "special" classes. Class 12 is for "Security Services" and Class 13, for example, is for "Public Utilities" and might be used for Smart City IoT applications. An AB flag is provided in MIB-NB. If it is set false, all devices are allowed to access. If the AB flag is set true, the device must read the SIB-14 before it attempts to access the network, which provides specific barring information for access classes. The IoT device needs to check whether its access class is allowed to connect to the network. If not, the device should try again at a later point of time or select a different PLMN, if available.

As mentioned in section "Coverage Extension," a device in bad coverage locations requires a significant number of repetitions to be configured for its upload traffic. During high network loads a PLMN might bar such devices and use available resources to serve more devices in good coverage locations instead. 3GPP Release 15 introduced a **coverage-level-specific barring** to exclude devices with a specific CE level (or worse). For this purpose, an RSRP threshold value is provided in SIB14-NB. If a device has measured RSRP below this threshold, it will be barred from this network.

As mentioned, applied subscription plan for an IoT project will determine which PLMNs should be used, i.e., usage cost will have significant impact on cell selection procedure. But if no priority PLMN is available or manual selection has been configured, the NB-IoT device will search for cells on appropriate frequency bands, reads the associated SIB information, selects a cell and measures the quality (i.e., noise) and the power level of the received NRS (NB Reference Signal). Relevant parameters for signal strength and signal quality are **RSRP (Reference Signal Received Power)** and RSRQ (Reference Signal Received Quality). RSRP is the measured power of the LTE reference signals spread across the broadband and narrowband portions of the spectrum. RSRP values, presented in dBm, are always negative, and the higher the number, i.e., the closer to zero it is, the higher the power of the signal. RSRP is a variation of parameter RSSI (Received Signal Strength Indicator) which is less relevant because it can be calculated with RSSI = RSPR/RSRQ.

In order to optimize cell selection procedure, local measurement of the actual signal strength at point of deployment should be part of the installation procedure for each IoT device. Then, measured RSRP value is compared to cell-specific threshold provided by SIB2-NB. For example, SIB2 might contain the following line:

```
pdsch-Config =
    referenceSignalPower : 15 dBm
```

In this case, the cell reference signal has been transmitted by eNB base station with 15 dBm output power. Let us assume that measured RSRP value at device target location is "−80 dBm." Based on these numbers you can calculate path loss between cell and device: Path Loss = (**referenceSignalPower**) − (RSRP measured by user device) = 15 − (−80) = 95 dBm. In this case, the device can consider itself to be in good range (see Fig. 9) and may decide to camp on this cell. In any other case or if other information provided by MIB/SIBs do not satisfy device requirements, the modem can look for other cells in reach, rank them, and provide a list of suitable options to the user device for consideration.

Mentioned second parameter RSRQ is defined by the LTE specification as the ratio of the carrier power to the interference power: essentially this is a signal-to-noise ratio measured using a standard signal. A connection with a high RSRQ should be good, even if the RSRP is low: the modem is able to extract the information in the

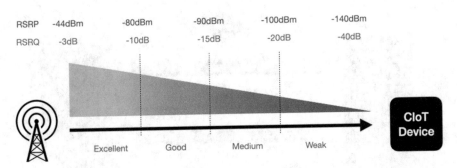

Fig. 9 RSRP and RSRQ—received reference signal power and quality

weak signal because of minimal noise. With RSRP and RSRQ values for all the nearby towers, the modem ranks usable cells for automatic or manual selection.

For IoT application developers using an off-the-shelf modem (see section "NB-IoT Cellular Network Modules" of chapter "Ingredients for NB-IoT Design Concepts") the good news is that they do not have to determine RSRP and RSRQ themselves. Instead, values are continuously being delivered by the modem, they can be read back by the application via proprietary AT commands (e.g., Quectel BG96 is using **AT+QCSQ**). In fact, different modem manufacturers might measure different RSRP and RSRQ values and score them differently, maybe because of their particular product ability how well it can extract signal. Figure 9 provides a rough indication of ranges for RSRP and RSRQ values to be measured at a CIoT device location within cell coverage area:

In manual cell selection mode, the user device host CPU can use these measurement results as an input for independent decision and submit corresponding AT commands in order to tell the modem which cell to camp on.

Radio Resource Control (RRC) and Random Access (RA) Procedures

After installation of an IoT device resp. first power up or relocation from an uncovered area, and a suitable cell has been identified, the device will have to establish a one-to-one communication channel with the eNB base station. Same applies when a user device wakes from power save mode (PSM). For this purpose, it performs a Random Access (RA) procedure to attach to the base station. This establishes a Radio Resource Control (RRC) connection to the base station, allocates resources and schedules for uplink data transmission.

For the radio interface two states are available: **RRC_Idle** and **RRC_Connected** (see Fig. 13). RRC_Connected state is required to receive or transmit data from or to the network. By nature, device power consumption in RRC_Connected state will be significantly higher than in RRC_Idle state, therefore it makes sense to keep

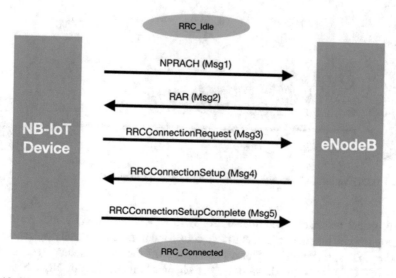

Fig. 10 RRC connection process (simplified)

RRC_Connected periods as short as possible. Once RRC_Connected state is established, only the eNB base station can release this connection and return device to RRC_Idle state. The user device cannot drop the RRC connection other than turning off the radio via AT command interface using the **AT+CFUN=0** command (see also section "Modem Interface"). The base station (eNodeB) has an "inactivity" timer for each module, and if no messages have been exchanged the base station it will release the connection and return RRC status to RRC_Idle. This inactivity timeout usually is configured for expiration after a couple of seconds (e.g., 10 s).

Similar to LTE user devices, NB-IoT devices have to request data transmission service from selected eNodeB base station using a random access (RA) procedure, which is a 5-message handshake (Fig. 10). First, an anonymous signal sequence, called preamble, is sent by the requesting device. The receiving base station then confirms detection of preamble in a message called RA reply (RAR). After RAR reception the device will transmit its unique connection request in Message 3. If Message 3 has been correctly received, base station then replies with Message 4, allowing the respective device to conduct data transmission and to apply RRC_Connected state.

For initial access to an LTE network, dedicated time-frequency resources called Physical Random Access CHannel (PRACH) are allocated periodically by eNB base station. **NPRACH** for NB-IoT is a slightly modified PRACH version in order to address NB-IoT specific characteristics. In fact, NB-IoT networks have to offer an efficient NPRACH procedure to support challenging NB-IoT objectives to manage

- A massive number of ten-thousands of devices covered by a single cell and to
- Access to devices located in areas with weak coverage.

This also means that NPRACH might react to actual service demand and/or to consider cell-specific characteristics of covered area (e.g., increased path loss due to buildings, trees, radio noise), and user devices will have to follow instructions provided by the cell they want to attach to.

The NRACH procedure always starts with a so-called preamble which is indicating an allocation request of a user device to connect. Mentioned SIB2 is periodically broadcasted by the cell and provides NPRACH configuration information which are common to all user devices. Relevant parameters are obtained by the requesting device and used to create an unpredictable NPRACH preamble. Preamble format, max. number of attempts, number of repetitions, etc. of NPRACH are set by requesting device according parameters provided by periodically broadcasted SIB2. SIB2 contains to dedicated random access information blocks called **RACH-ConfigCommon-NB** and **NPRACH-ConfigSIB-NB**. **RACH-ConfigCommon-NB** is used to specify the generic random access parameters, whereas **NPRACH-ConfigSIB-NB** is used to specify NPRACH configuration for the anchor and non-anchor carriers, for example:

For preparation to start random access procedure (incl. further attempts and repetitions), the device should have a full set of RACH parameters available (see [10], Sect. 5 "Random Access Procedure").

The requesting device needs to determine an appropriate NPRACH configuration according to its coverage class estimation. In general, each network cell specifies its own thresholds for coverage enhancement (CE) level assignment. As mentioned before (see section "Network Selection"), the user device can determine its CE level via measurement of RSRP signal strength. If supported, the cell can specify up to three RSRP thresholds (see **rsrp-ThresholdsPrachInfoList** in Fig. 11) corresponding to four CE levels (i.e., CE0 < Threshold 1 < CE1 < Threshold 2 < CE2 < Threshold 3 < CE3) to be selected by the device according to its coverage class. SIB2 will contain three NPRACH configurations for four CE levels. If the network does not specify any RSRP thresholds, only one single NPRACH configuration is supported, so all devices will have to use it regardless of their actual path loss to the serving base station.

The base station allocates distinct NPRACH for each CE level which occur periodically. To serve up to four NPRACH for the four CE levels, base station divides the 180 kHz bandwidth into 48 subcarriers, each with a subcarrier spacing of 3.75 kHz. The basic unit of the subcarrier allocation for an NPRACH of one CE level is 12 subcarriers. Therefore, an NPRACH for a CE level can have 12, 24, 36, or 48 subcarriers [9]. The requesting device starts a random access procedure by transmitting a preamble in an NPRACH at its CE level which has been determined before. The NB-IoT sequence to be used for the preamble is the same for all devices. Each preamble consists of four symbol groups transmitted without gaps.

Frequency hopping is applied to these symbol groups, i.e., they are each transmitted on different subcarriers (out of allocated group of 12 subcarriers). Among the 12 possibilities, the user device selects the subcarrier for transmission of the first symbol group, if actual RA procedure has been initiated by the device itself. Then, subcarrier selection for next symbol groups is pseudo-random based on an algorithm which has been designed in a way that applied hopping schemes will never overlap and maximize availability of congestion-free preambles (Fig. 12).

Parameter Name	Value options	Short Description	3GPP Reference
NPRACH-ConfigSIB-NB			
nprach-PreambleFormat	fmt0, fmt1, etc.	TDD preamble format	TS 36.211 [14], clause 10.1.6
nprach-Periodicity	40, 80, 160, 240, 320, 640, 1280, 2560 ms	NPRACH resource periodicity	TS 36.211 [14], clause 10.1.6
numRepetitionsPerPreambleAttempt	1, 2, 4, 8, 16, 32, 64, 128	Number of NPRACH repetitions per attempt for each NPRACH resource	TS 36.211 [14], clause 10.1.6
nprach-SubcarrierOffset	0, 12, 24, 36, 2, 18, 34 In number of subcarriers, i.e. offset from subcarrier 0	frequency location of the NPRACH resource.	TS 36.211 [14], clause 10.1.6
nprach-NumSubcarriers	12, 24, 36, 48	Number of sub-carriers allocated to a NPRACH resource	TS 36.211 [14], clause 10.1.6
nprach-NumCBRA-StartSubcarriers	8, 10, 11, 12, 20, 22, 23, 24, 32, 34, 35, 36, 40, 44, 48	Number of start subcarriers from which a UE can randomly select a start subcarrier.	TS 36.321 [15]
nprach-StartTime	8, 16, 32, 64, 128, 256, 512, 1024 milliseconds	Start time of the NPRACH resource in one period	TS 36.211 [14], clause 10.1.6
maxNumPreambleAttemptCE	3, 4, 5, 6, 7, 8, 10	Maximum number of preamble transmission attempts per NPRACH resource	TS 36.321 [15]
npdcch-NumRepetitions-RA	1, 2, 4, 8, 16, 32, 64, 128, 256, 512, 1024, 2048	Maximum number of repetitions for NPDCCH common search space (CSS) for RAR, Msg3 retransmission and Msg4	TS 36.321 [15], clause 16.6.
npdcch-StartSF-CSS-RA	1.5, 2, 4, 8, 16, 32, 48, 64	Starting subframe configuration for NPDCCH common search space (CSS), including RAR, Msg3 retransmission, and Msg4	TS 36.321 [15], clause 16.6.
npdcch-Offset-RA	zero, oneEighth, oneFourth, threeEighth	Fractional period offset of starting subframe for NPDCCH common search space	TS 36.321 [15]
rsrp-ThresholdsPrachInfoList	one or two threshold values in dBm	CE level criterion for user device to select a NPRACH resource	TS 36.321 [15]
RACH-ConfigCommon-NB			
powerRampingStep powerRampingStepCE	e.g. 2 dB	Power ramping step	TS 36.213 [20] and TS 36.321[15]
preambleInitialReceivedTargetPower	e.g. -104 dBm	PRACH power that eNB expects to receive	TS 36.321 [15]

Fig. 11 NPRACH parameters (selection)

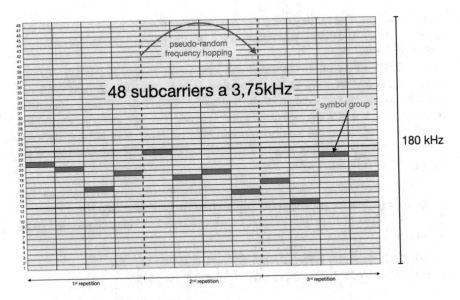

Fig. 12 Preamble sequence (example)

Each symbol group contains one cyclic prefix CP plus five identical symbols, CP length depends one of two available RACH formats determined by SIB2 before. For a transmission attempt, the same preamble is repeatedly transmitted within the same NPRACH (see parameter **nprach-Periodicity** in Fig. 11). The first repetition is transmitted at a subcarrier the user device has chosen from the list of subcarriers allocated to the CE level. For each cell a suitable number of repetitions (up to 128) for different CE level is pre-configured by SIB2 in order to ensure proper reception (see parameter **numRepetitionsPerPreambleAttempt** in Fig. 11). Consequently, NPRACH of the higher CE level has longer duration than that of the lower one. Each time the preamble transmission counter is reaching this parameter, applied CE level is increased. Thus, after the configured number of reattempts in the initially selected CE level, reattempts are performed using NPRACH for a higher CE level—with a higher number of repetitions. If the base station detects an NPRACH preamble, it returns a random access response (RAR), also known as Message 2 (see Fig. 10).

The RAR contains timing and uplink resource allocation as an input to the requesting device for preparation of its RRCConnectionRequest (Message 3). In Message 3 the device will include its identity and indicate its bandwidth demand as well as its capability to support for multi-tone traffic and multi-carrier support. In addition, the device will report its power headroom resp. applied power for preamble transmission P_{NPRACH} vs. maximum transmit power P_{MAX}.

Collision occurs if two or more devices transmitting at the same NPRACH choose the same initial subcarrier. In case of collision, the requesting device has to back off, select a new initial subcarrier and retransmit the preamble in next available NPRACH. This process repeats until the maximum number of attempts in this CE level is reached. The failed device can restart the whole RA procedure in the next higher CE level. The device will declare a RA failure if the RA procedure fails at the highest CE level.

Probability of preamble collisions will increase in congested NB-IoT environments where lots of devices are requesting service. But independent from cell capacity, the individual location of the requesting NB-IoT device resp. its distance to the serving cell is a key aspect for good network coverage and a successful RA process. By definition, the maximum transmit power P_{MAX} of an NB-IoT device is 23 dBm. Other parameters required for calculation of required transmit power P_{NPRACH} are as follows:

1. eNodeB base station receiver sensitivity resp. expected signal strength specified by SIB2 parameter **preambleInitialReceivedTargetPower** (see Fig. 11).
2. Coupling loss of data transmission between NB-IoT device and eNodeB (path loss) with

   (a) SIB2 parameter **referenceSignalPower** (see Fig. 11) plus
   (b) Strength of RSRP reference signal at device location

Example

Let us assume, for example, these input values for P_{NPRACH} calculation:

- **preambleInitialReceivedTargetPower**: −104 dBm
- **referenceSignalPower** = 18 dBm
- RSRP measured by user device: −88 dBm

Calculation:

P_{NPRACH} = **preambleInitialReceivedTargetPower** + Path Loss
= −104 + 106
= 2 dBm

In this case, applied P_{MAX} of 23 dBm provides sufficient headroom to ensure a seamless random access procedure at lowest coverage enhancement level CE0.

Finally, in Message 4 the network resolves any contention resulting from multiple devices transmitting the same initial RA preamble and provides a connection setup. After reception of Message 4 the device will transit from RRC_Idle state to RRC_Connected mode. The first message of the device in RRC_Connected state is Message 5 RRCConnectionSetupComplete (see Fig. 10).

Later, when a device has returned to the RRC_IDLE state, it may either use again the random access procedure if it has new IoT data to send, or waits until it gets paged. Another option for the device is to store a copy of relevant SIB data and use it after cell reselection or return from out of coverage.

Power Saving Methods

Enabling a significant reduction of overall device power consumption was a major NB-IoT objective. Goal was to enable battery-powered devices with a lifetime of more than 10 years. In fact, besides support for a large number of connected devices (network capacity) and wide/deep network coverage, NB-IoT provides low power modes and power saving mechanisms are key selling points for IoT applications using simple zero-touch and maintenance-free devices a long product lifetime which is not limited to a short battery lifetime.

Of course, promised 10 years battery lifetime will not work for all kind of CIoT devices. Instead, focus is on typical IoT applications which do not require permanent and immediate device availability, e.g., a fluid level meter which is supposed to transmit consumption data only once per day. This kind of IoT device will transmit small amount of data periodically, and afterwards it will stay on hold or idle for the rest of the day. Refer to section "NB-IoT Use Cases" of chapter "IoT Target Applications" for further target applications.

3GPP standardization work has been driven by these IoT-typical characteristics to support implementation an efficient power management strategy for NB-IoT devices. For this purpose, PSM and eDRX power saving modes and other mechanisms are offered to minimize active radio time between the IoT device and the network.

Power Saving Mode (PSM)

PSM is a power saving feature designed for NB-IoT devices which allow to reduce battery power to a minimum. In fact, PSM will switch off the device radio and most part of the network interface which reduces power consumption of this circuit to a few microamps.

But, as usual when leaving RRC_Connected, also during PSM period the IoT device still remains registered in the network, meaning that the device closes the AS connection yet keeps the AS/NAS status, i.e., higher layer connection configuration. Consequently, when the device wants to leave PSM it will just have to resume the previous connection without having to re-attach or re-establish the PDN connections. This avoids extra power consumption due to additional communication with the base station for the higher layer connection establishment procedure.

However, designers will have to keep in mind that—as long as in PSM mode—the device will be unreachable because radio is off, no paging notifications will be received. There are two ways how to wake up the device from PSM state:

1. Event originated from IoT device (e.g., from an integrated sensor) or
2. Expiration of network timer T3412 (TAU or Extended Timer)

T3412 is configured in a way that the user device will wake up periodically to perform a Tracking Area Update (TAU). TAU is a standard LTE procedure used by the device to notify periodically its availability to the network. TAU timer T3412 is similar to LTE but can be programmed with a longer expiration time leading to extremely infrequent periodic TAU events of up to 413 days (!). Expiration of T3412 will end RRC_Idle state and initiate a new or resumed connection.

Besides TAU, in RRC_Connected state all data transmissions between user device and eNB base station will be performed. During this period, the user device

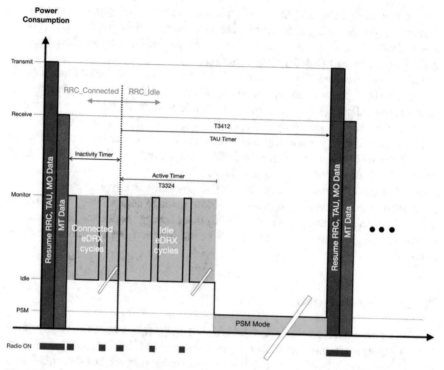

Fig. 13 Low-power modes and timing

listens to paging occasions from the network and will receive queued downlink data (mobile terminated—MT data), if any. And it will perform pending uplink transmissions (mobile originated—MO data), if any. Figure 13 is outlining different operational scenarios of an NB-IoT device and corresponding power consumption levels. Note that power consumption levels correspond to radio activity for different modem activities: transmit data, receive data, listening to paging occasions. For low-power NB-IoT device design it will be crucial to keep these active periods as short as possible, see extra section "Low Power Device Design" of chapter "Designing an NB-IoT Device."

Extended Discontinuous Reception (eDRX)

Standard LTE power saving feature **DRX** (DRX = discontinuous reception) is designed for efficient support of downlink communication. It can be executed when device is in RRC_Idle mode. In RRC_Idle state, the IoT device cannot transmit data or request resources from the network, but the so-called NPDCCH channel is tracked to determine if there is downlink data pending. During this paging period, energy saving is achieved due to the fact that only some of the subframes are monitored, i.e., the IoT device alternates between active listening of a paging occasion

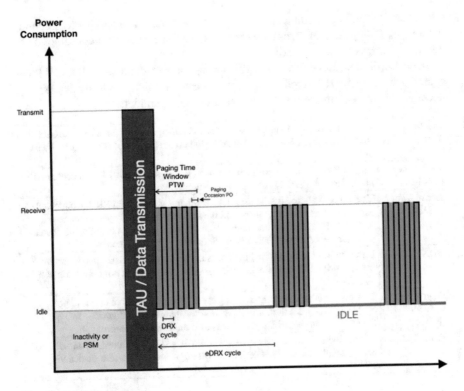

Fig. 14 eDRX timing details

(PO) and sleep for the following paging occasions. This "discontinuous reception" will result in extra latency, but DRX cycle length can be configured by the user in a way that it will not have any negative impact on the IoT application. In RRC_Idle state, DRX cycles of 128, 256, 512, and 1024 radio-frames are supported, i.e., from 1.28 to 10.24 s with a radio-frame of 10 ms each.

The **Extended DRX (eDRX)** feature goes one step further and extends duration of DRX cycles to allow a device to stay in low-power mode after paging window has closed (see Fig. 14), i.e., the IoT device will not listen to control channel for paging messages resp. download traffic until the configured eDRX cycle has ended. If supported by the network, length of this eDRX period is determined by timer T3324 which is programmable by multiples of a hyper-frame (1024 radio-frames = 10.24 s) which results in a maximum duration of ca. 3 h. From IoT device perspective, eDRX is initiated using a dedicated command via AT command interface of the network module (see section "Modem Interface"). Attached network must be able to handle temporary device unavailability and cache network requests until next paging period in RRC_Connected state occurs. This means that requested timing parameters are subject to negotiation, i.e., not confirmed by network yet. If accepted, network will return configured PTW and eDRX cycle values to the requesting device. Based on these agreed timings, a latency-tolerant IoT application can check for pending commands, for example, every hour and apply a corresponding long eDRX cycle afterwards in order to decrease average power consumption to a few milliamps during this period.

Involved timers can be used to adjust timing and schedules of eDRX and PSW cycles according to IoT application requirements, e.g., in terms of availability, latency. This is a summary of relevant conditions and parameters:

Inactivity Timer	Expiration of this timer causes user device change state from RRC_Connected to RRC_Idle. This timer is controlled by the eNodeB and **not** configurable by user device.
TAU Timer (T3412)	This network timer determines after which time a periodic TAU procedure will be performed. Period is user device configurable.
Active Timer (T3324)	This timer is started by the user device when it moves from RRC_Connected to RRC_Idle state. During this period the device runs eDRX cycles, i.e., is reachable for downlink communication. As soon as T3324 expires, the device will enter PSM mode. T3324 is configurable by user device.
DRX cycle	Duration is a multiple of paging occasion (PO) cycles of 1028 ms. User device listens during PO, then sleeps until next PO. This timer is **not** configurable by user device.
Paging Time window (PTW)	This period is representing a paging event, it is composed of several DRX cycles. DRX cycles fit in this PTW window, length is configurable by user device, i.e., number of DRX cycles is also under device control.
PSM mode	PSM is the ultimate device low-power mode, radio is switched off and device is not reachable during PSM period. It is configurable by the user device via T3324.

Based on these configurable parameters, NB-IoT offers are offering a versatile toolbox for application developers which allows adjustment of applied schedules and timings according to specific needs. In particular, use of implemented power saving mechanisms for a long battery life can be balanced with other use case requirements, e.g., maximum allowed latency.

Release Assistance Indication (RAI)

Release assistance indication (RAI) allows the NB-IoT device to indicate to the eNB during RRC_Connected mode that it has no more UL data and that it does not anticipate receiving further DL data. In this case, the device would like the network to release the device to RRC_Idle mode and quickly to save radio power, release assistance indication (RAI) is introduced in Release 13 for Non-Access Stratum (NAS) and in Release 14 for Access Stratum (AS). When AS RAI is configured, UE may trigger a buffer status report (BSR) with size of zero byte as a request to eNB for an **early transition from RRC_Connected to RRC_Idle state**. Without RAI, by default the device would have to wait for the eNB to release the connection via explicit signaling because the eNB is not fully aware of the UE data buffer or the expected DL traffic. Thus, it sends a connection release message that forces the UE to enter idle state after the RRC inactivity timer expires.

RRC Suspend/Resume

This method for signaling reductions allows an NB-IoT device to resume a connection which was previously suspended. As a prerequisite it requires an initial RRC connection establishment (e.g., a periodic TAU) that configures the radio bearers and the Access Stratum (AS) security context incl. security configuration for data protection (e.g., encryptions keys) for an NB-IoT device in the network. Then, the NB-IoT device can enable the RRC connection to be suspended and resumed whenever needed, if the device supports this feature and has been configured accordingly.

When the NB-IoT device transits to RRC Idle state, the RRC Resume procedure will store the connection context and assign a resume ID. This ID is signaled by the RRCConnectionRelease message from the network. Later, when there is new traffic, the device can submit RRCConnectionResumeRequest message containing the associated resume ID to be used by the eNB to access the stored context. On top of resuming a prior connection, the RRC Resume procedure also allows to transmit a small amount of uplink data in Message 5 RRCConnectionSetupComplete (see Fig. 10). For this purpose, payload data is multiplexed with RRC packet data in Message 5 [12].

Preserving the device context instead of releasing it each time it returns to RRC_ Idle state means for the device to **skip AS security setup and RRC reconfiguration** for each data transfer, saving a considerable signaling overhead.

Early Data Transmission (EDT)

EDT was introduced in 3GPP Rel 15 and advances benefits of RRC resume procedure which is available also for Rel 13 and Rel 14 NB-IoT devices. EDT allows a device to exchange network data early during random access (RA) procedure in Message 3 resp. Message 4. If EDT is being used during RA procedure, the device can complete its **transmission of user data in RRC_Idle mode** and does not have to transit to RRC_Connected state at all. However, EDT is limited to small data payloads. Uplink EDT supports transport blocks sizes (TBS) of 328 up to 1000 bits (Fig. 15).

Wake-Up Signal (WUS)

In RRC_Idle mode during eDRX cycles the NB-IoT will have to check periodically for paging messages from the network. For this purpose, the device has to activate its baseband receiver and monitor downlink channels NPDCCH (control) and NPDSCH (payload). Unfortunately, at most possible paging occasions (PO) there will be no messages addressed the device, e.g., at nighttime or because unplanned download traffic does not apply to target IoT use case at all. This means that applied radio power for monitoring NPDCCH channel for has been wasted in these cases and reduce battery lifetime without need. In order to increase paging efficiency, a

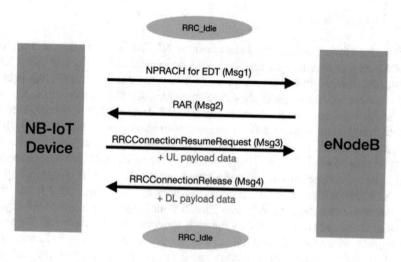

Fig. 15 Early data transmission (EDT)

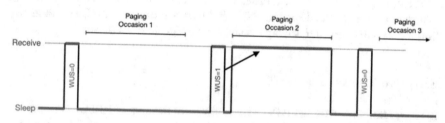

Fig. 16 WUS feature

wake-up signal (WUS resp. NWUS for NB-IoT) was introduced in 3GPP Rel 15. The idea is to **indicate relevance of the following paging occasion** to the NB-IoT device by a single-bit WUS information. WUS takes less time to transmit and is much easier to detect compared to required demodulation and decoding efforts for the full paging event. As a consequence, the NB-IoT device will monitor the NPDCCH channel only in case an WUS signal has been detected before. If no WUS was not set, the device can skip the next PO and stay in power save mode for more time.

Figure 16 illustrates how WUS signal works. Prior to each paging occasion, WUS informs the NB-IoT device the associated paging occasion will indicate pending download traffic for the device or not. If not, i.e., in periods 1 and 3 with WUS = 0, the device will return to sleep immediately. In period 2 with WUS = 1 the device has been alerted and will be ready to monitor and decode NPDCCH channel data accordingly. With the help of WUS, the device can reduce reception periods and save unnecessary power consumption for paging occasions which are not addressed to the IoT device anyway.

WUS is an optional Rel 15 feature requiring support of the device modem and the network as well. On device side, WUS potentially can unfold extra power efficiency if a modem offers a **dedicated low-power WUS detector** to wake up the main baseband receiver in case an WUS was detected. Such a WUS receiver piece of integrated hardware will allow to keep the main receiver switched of during paging as long as the IoT device itself is not addressed [12–14].

Latency

In general, NB-IoT is aiming at latency-tolerant applications and use cases which do not require 24/7 online availability for external parties and/or immediate reaction of the device in case of downlink requests. But nevertheless, one of the NB-IoT objectives was to keep latency for uplink traffic below 10 s. This means that—even under worst case conditions (i.e., MCL = −164 dBm)—the delay to transmit a small data package to the network should not exceed 10 s for a connected NB-IoT device.

During PSM periods, a NB-IoT device is not reachable and first has to wake up before it can transmit any uplink data. This an example how use of power saving features is impacting latency and it shows that both requirements need to be balanced carefully. In addition, NB-IoT devices in poor device coverage conditions assigned with CE1 or CE2 coverage enhancement levels have to repeat message transmissions multiple times, this will reduce data rate and increase latency.

IoT device settings can be controlled with AT commands (see section "AT Command Interface" of chapter "Ingredients for NB-IoT Design Concepts") and some network settings are public (e.g., T3412 TAU Timer), but others are not. For example, an NB-IoT cannot control duration of Inactivity Timer or settings of coverage enhancement thresholds or used TX output power. As a consequence, a network user cannot easily estimate network latency and transmission time. Instead, some simulations and real-world measurements are available to provide more evidence. For example, network infrastructure specialist Ericsson has published some uplink latency numbers [1] for NB-IoT devices at different cell locations at cell border (+0 dB) and beyond (+10 dB and +20 dB):

Coverage (dB)	Sync (ms)	Sys Info MIB/ SIB (ms)	PRACH (ms)	UL transmission (ms)	max. Uplink Latency (s)
+0	340	151	946	167	1604
+10	340	151	1396	1158	3085
+20	520	631	2500	3972	7623

Another paper [15, Table 1] is summarizing best and worst case values for these timing elements. For example, transmission of the PRACH preamble can take 5.6 ms minimum (Format 0, 1 repetition) or 819.2 ms maximum (Format 1, 128 repetitions).

Another report [12, 16] about uplink latency of an 84-byte data package also looks at different NB-IoT operation modes (stand-alone, guard-band, in-band):

Coverage (dB)	Stand-alone (s)	Guard-band (s)	In-band (s)
+0	0.3	0.3	0.3
+10	0.7	0.9	1.1
+20	5.1	8.0	8.1

In this study, transmission time with a stand-alone deployment was shorter compared to in-band or guard-band operation modes because full radio power was allocated to the NB-IoT carrier. More transmission power leads to better coverage and less repetitions. This is speeding up transfer of the final downlink acknowledgement message from the network to the transmitting IoT device. This advantage has minor impact on devices in good coverage conditions because no repeated transmissions are slowing down the overall process. At a coupling loss of less than 144 dB, latency is mostly determined by the time needed to access the network (incl. random access procedure) and to acquire system configuration blocks (MIB, SIB). But at maximum coupling loss 164 dB, latency is dominated by signal repetitions used for coverage extension.

Very obviously, NB-IoT users are facing a significant range of latency effects, mainly caused by assignment of coverage enhancement levels CE0–CE2. For a typical NB-IoT target application (see section "NB-IoT Use Cases" of chapter "IoT Target Applications") it will not matter too much if device-originated IoT data will arrive a few seconds later. But for battery-powered devices, long data transmission periods are critical because they have significant impact on lifetime. For further information about low-power design aspects and how to calculate long-term current consumption see section "Low Power Device Design" chapter "Designing an NB-IoT Device."

Ingredients for NB-IoT Design Concepts

"IoT" is an umbrella term for different applications with certain similarities, e.g., they are all connected and perform some kind of edge processing. But most IoT applications require use-case-specific, customized IoT devices because they are part of a unique solution contributing to the overall approach how a business case is being addressed.

Different requirements might apply in terms of sensor functions, power supply, computing speed, cost, packaging, security, etc. But customization on hardware level very often comes down to selecting suitable components for remote sensing and acting, i.e., standard products will do the job and no dedicated custom circuits will be required. The core IoT device hardware platform is quite the same for all: you need a modem, a general-purpose application processor (MCU), some memory for code/data storage and standard interfaces like I²C, SPI, or GPIOs to connect selected peripheral components.

A universal IoT hardware platform will offer the following building blocks, see Fig. 1:

1. **Cellular network interface** incl. modem, antenna, SIM card interface
2. **IoT application processor** (MCU) incl. memory
3. **IoT peripherals** like a sensor or an actuator

Due to the huge business potential of IoT applications, many suppliers of IoT components and services are struggling for customer attention. **"One-stop-shopping"** is a common industry trend for suppliers to offer IoT solutions rather than single products or resources, e.g., hardware/software bundles, development kits and tools, ready-to-use software stacks, cloud support, connectivity services, etc. This trend is beneficial for application developers who can minimize design risk and turn-around-time. Another trend is **online support** allowing application developers to access useful resources and interact with experts at anytime from anywhere.

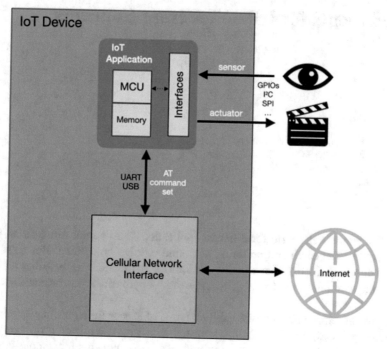

Fig. 1 Block diagram of a simple generic IoT device

NB-IoT Cellular Network Modules

In general, interfacing to a cellular network is complex and requires advanced RF and analog design expertise. On top of this, application developers expect a certain level of abstraction from complex 3GPP standards resp. from low-level knowledge of NB-IoT physical layer, device-network synchronization, random-access procedures, etc. Know-how on this level is useful but not required for IoT application development. Instead, for efficient work, higher-level functions (API) and efficient tools are needed. Cellular modem vendors have recognized an increasing IoT demand from different industry segments, so they started to leverage their modem expertise for their offer of **comprehensive and easy-to-use subsystems**. These cellular network modules are **for application developers** requiring cellular connectivity for their project without spending too much time with underlying cellular network technology itself (Fig. 2).

In fact, core of each cellular network module is a **modem** (modulator-demodulator), i.e., a data converter which is modulating a carrier wave to encode digital data for transmission. In our case, transmission medium is a wireless cellular NB-IoT network with carrier frequencies of up to 2 GHz and output transmit power of up to 23 dBm resp. 200 mW. This mix of digital, analog, and power requirements means extra challenge for integration within a single semiconductor product. Thus, cellular network modules are usually containing a mixed-signal modem chip plus

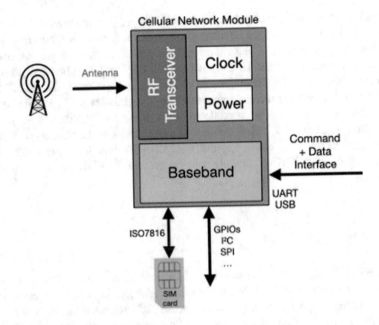

Fig. 2 Cellular network module

extra power amplifier and some other discrete components altogether in a compact multi-chip SMD package (e.g., a 96-pin LGA with 16 × 26 × 2.4 mm).

A typical cellular network module contains

- Modem incl. command/data interface to IoT application (UART or USB)
- RF interface, amplifiers, filters
- Clock generation and distribution
- Power management
- SIM card interface
- Microcontroller, OS, firmware, memory
- Analog/Digital Converter
- Peripheral interfaces (GPIOs, I²C, SPI, etc.)

Most of the modem work is digital data processing on baseband level, i.e., it is a specialized embedded computing device. Usually, a general-purpose CPU core (e.g., by ARM) is used, controlled by a flash-based dedicated modem firmware. This firmware is subject to occasional updates in an effort to follow NB-IoT extensions specified in yearly 3GPP releases, for added features, bug fixes, etc. For deployed IoT devices which are prepared for it, firmware can be upgraded "over-the-air" (OTA), i.e., using an NB-IoT connection for transmission of the new firmware version.

As a matter of fact, a cellular network module for NB-IoT (short: "NB-IoT module") is the most important component of a NB-IoT device, mainly because of the complexity of the LTE/NB-IoT standard and corresponding infrastructure. All other components of an IoT device are built "around" the NB-IoT module, even though it is controlled via AT command interface by an IoT application program running on an extra MCU (see Fig. 1). For example, in a typical device design scenario the modem will wake up the rest of the design when PSM state has been ended by the network. Main reason for the major role of the NB-IoT module is the overall complexity of 3GPP cellular technologies requiring a certain level of dedicated expertise which is offered by all module vendors mentioned below. From an IoT application point of view, NB-IoT modules unfold most of the benefits of NB-IoT cellular connectivity by delivering essential network functions, protocols, tools, support. Price of a NB-IoT module will be around 20 EUR per unit for small quantities.

Vendor Overview

Major NB-IoT module vendors are (in alphabetical order): Nordic Semiconductor, Quectel, Sierra Wireless, Telit, and u-blox. Interestingly, all of these companies are wireless network specialists, no broadline semiconductor manufacturers like Texas Instruments or STMicroelectronics or Samsung are offering any cellular modem chips.

- **Nordic Semiconductor**, https://www.nordicsemi.com

    - Nordic Semiconductor is a Norwegian semiconductor company founded in 1983. Nordic is specializing in wireless communication technology for IoT (Bluetooth, ANT+ Thread, Zigbee, WiFi, LTE-M/NB-IoT). Listed at Oslo Stock Exchange. 2020 revenues: 405 mio USD, 1000 employees.

- **Quectel**, https://www.quectel.com/product-category/lpwa-modules/

    - Quectel entered business 2020 in Shenzhen/China with a GSM/GPRS module and exclusively focused on cellular network modules. Claims to be the "world's largest and fastest-growing supplier of IoT modules." Listed at Shanghai Stock Exchange since 2019. 2020 revenues: 935 mio USD, 2300+ R&D engineers.

- **Sierra Wireless**, https://www.sierrawireless.com/products-and-solutions/embedded-solutions/

    - Sierra Wireless entered business in 1997 with an embedded cellular module. Focused on cellular modules and services. Headquartered in Canada, listed at NASDAQ. 2020 revenues: 448 USD, 1000+ employees.

- **Telit**, https://www.telit.com/m2m-iot-products/cellular-modules/

    - Telit started in 1997. Cellular IoT, WiFi, Bluetooth, GPS/GNSS products, solutions and services. Listed at London Stock Exchange. 2020 revenues: 343 mio GBP.

- **u-blox**, https://www.u-blox.com/en/cellular-modules

  - u-blox started as a spin-off from the Swiss Federal Institute of Technology in 1997, first product was a GPS receiver. Listed at Swiss stock exchange since 2007, they still specialize on IoT solutions for cellular networks and GPS/GNSS. 2020 revenues: 333 mio CHF, 1200+ employees.

Figure 3 provides an overview of typical NB-IoT modules available at time of writing. A comprehensive competitive comparison would have to include several additional technical criteria (e.g., operating temperature, supply voltage) and cost. Instead, our table is taking a top-level view focusing on key product differentiators from a device design perspective, and it is highlighting some power saving or security features, if any. It also includes some "soft" criteria like online support.

Manufacturer		Nordic	Quectel	Sierra Wireless	Telit	u-blox
Type		nRF9160	BG95 Series	WP7700	ME910x	SARA-R5 Series
Standards		NB1, NB2, M	NB1, NB2, M	NB1, M	NB1, NB2, M	NB1, NB2, M
3GPP Release		14	14	13	14	14
GPRS fall-back?		no	yes	yes	yes	no
GPS option ?		yes (int.)	yes (int.)	yes (int.)	yes (int.)	yes (int.)
open OS ?		no	no	yes	no	no
embedded user application ?		yes	no	yes	yes (IoT AppZone)	no
security hardware		ARM CryptoCell				secure element CC EAL5+
Power Consumption	Power Class 6 (14dBm) ?	no	no	no	no	yes
	antenna tuning control interface	yes	no	optional	no	yes
	PSM [µA]	2,7	3,9	22	3	0,5
wake-up	Input		PWRKEY	POWER_ON	WAKE	PWR_ON
TX indicator	Output			TX_ON		
PSM indicator	Output				PWRMON	V_INT
Online support	private support request				qualification req'd	yes
	Community	yes	yes	yes	no	yes

last update: May 2021

Fig. 3 NB-IoT modules overview

All mentioned modules are compliant with 3GPP Releases 13/14 only, i.e., they are not covering subsequent specifications. This means that efficient Rel. 15 "Early Data Transmission (EDT)" power saver feature is not available for NB-IoT device designs yet. And Rel. 14 **power class 6** (14 dBm) capability for reduced transmission power of NB-IoT devices is offered only by u-blox SARA-R5 so far. In fact, NB-IoT devices can benefit from power class 6 only if they are operated in a compliant network infrastructure, so we might have kind of a chicken-and-egg problem here.

Independently from applied level of transmission power, antenna, and transmitter impedances should be matched properly. This does not cost much, but helps to improve efficiency and quality of uplink data channel, see section "Low Power Device Design-Matching of Antenna" of chapter "Designing an NB-IoT Device." For this purpose, a matching C/L network must be inserted in between transmitter and antenna— depending on actual NB-IoT frequency band. Some modules have implemented corresponding control software and AT commands as modem firmware out-of-the-box (Nordic nRF9160, u-blox SARA-R5). For others, the user IoT application program will have to take care and handle activation of an **antenna matching** network.

Another power saver is **PSM input current** which makes a significant difference in terms of product lifetime, esp. for battery-powered NB-IoT devices. With just 0.5 µA u-blox SARA-R5 is best-in-class in this particular field. First of all, PSM is an NB-IoT feature which is applied to reduce modem power consumption during inactivity periods, but it can also be used to trigger other device component to enter sleep mode. For this purpose, some modules offer an **PSM indicator pin** to external elements that the modem is currently in PSM state (u-blox SARA-R5, Telit ME910x).

For "Remote Monitoring/Detection" IoT use cases, a path in opposite direction might be required: a sensor might want to indicate a wake-up event to the NB-IoT modem in order to terminate PSM mode and trigger uplink data transmission. For this purpose, most modules are offering a dedicated input pin (called PWR_ON or WAKE or similar) and associated embedded function.

For IoT deployments in areas with uncertain NB-IoT coverage, a **GSM/GPRS fallback option** can help. In fact, GPRS was the first cellular data service back in year 2000 and still supported by most networks. GPRS is available almost everywhere in every part of world, see https://www.gsma.com/coverage/. Some manufacturers offer GPRS as an integrated additional feature of an NB-IoT module, or as a replacement option with same pinout (Quectel, Sierra Wireless, Telit).

For some IoT applications (see section "Object Tracking/Localization" of chapter "IoT Target Applications") require precise localization of an IoT device via satellite (GPS, GNSS, etc.). In fact, many manufacturers of cellular network modules also have a product line for **satellite positioning**, so they offer some kind of CIoT/ GPS bundle which is combining both technologies. In fact, all vendors listed in Fig. 3 have a single-chip solution with an integrated GPS engine and an AT command extension dedicated for GPS control and interaction.

In general, each network module comes with a handful of **GPIOs and standard interfaces** like I²C, which can be used by the IoT application. For this purpose, custom AT commands are offered which can be used to exchange data with IoT peripherals or for control of an antenna matching network, for example. On top of sharing I/O resources, some module manufacturers are even **sharing the module MCU** with the host IoT application which would normally be running on a separate external MCU, see Fig. 1. This is a strong feature which is reducing component count (bill-of-material) and PCB space. Nordic Semiconductor as well as Telit are offering dedicated development environments for this purpose. Sierra Wireless WP7700 offers an open source platform (https://legato.io) for IoT application development, module OS is a fully user-interactive Linux derivative.

End-to-end security for IoT data and protection against misuse of IoT applications are important aspects to be addressed by design. Some IoT applications are requiring an extra level of protection because failure would create damage or financial loss. In these cases, so-called **secure elements** resp. security certifications are required for an IoT device to qualify as a candidate for short-listing. Two NB-IoT modules (Nordic nRF9060 and u-blox SARA-R5) are offering dedicated security hardware. See section "End-to-End IoT Data Security" for further information.

Last but not least, an appropriate level **manufacturer online support** and services are particularly important for many small- and medium-scale IoT projects with no direct link to products experts and dedicated field application staff. See extra section "Suppliers and Online Support."

AT Command Interface

For modem control and data transfer between an IoT application program and a cellular network module, a special command language is being used (see Fig. 1). It is called **AT command set**. Usually, the IoT application program is running on a separate MCU communicating with the NB-IoT modem via UART or USB interface. But some integrated solutions are also available, see Fig. 3. AT commands are defined as part of 3GPP standard under 3GPP TS 27.007. That implies that all cellular network modules have to implement this API (Fig. 4).

The AT command set consists of a series of short text strings. AT commands always start with "AT" which is a mnemonic code for "Attention." We have four types of AT commands, namely Test, Read, Set, and Execute (Fig. 5).

- **Test**. The Test command is mainly used to check whether a command is supported or not by the modem.

  - syntax: **AT<command name>=?**

    Example:

    +CGMI=? (Request Manufacturer Identification)
    Response:
    OK

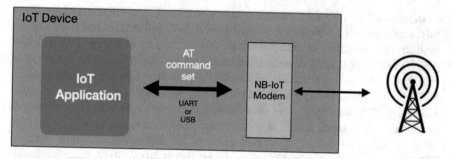

Fig. 4 AT command interface

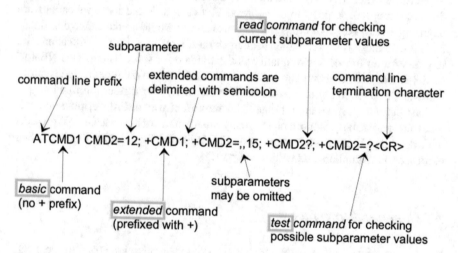

Fig. 5 AT command syntax

- **Read**. The Read command is mainly used to check the current setting of the modem parameter required for a specific operation.

  - syntax: **AT<command name>?**

    Example:

    AT+UBANDSEL? (check current LTE frequency bands)
    Response:

```
+UBANDSEL:  800,850,900,1800,1900,2100,2600
OK
```

- **Set**. The Set command is mainly used to modify settings of the modem required for a specific operation.

– syntax: **AT<command name>=value1, value2, ..., valueN**

Example:

AT+UBANDSEL=1800,2100,2600 (change the operating LTE bands)
response:

OK

- **Execution**. The Execution command is used to carry out an operation.

  – syntax: **AT<command name>=parameter1, parameter2, ..., parameterN**

  Example:

  AT+CMGS="67890"<CR> Hello world<Ctrl-Z>(send a text SMS)
  Response:
  +CMGS: 6
  OK

As a starting point, AT commands can be used to retrieve **general information** about the user device, e.g., manufacturer name, model number, firmware version, International Mobile Equipment Identity (IMEI) number, IMSI (International Mobile Subscriber Identity), ICCID (serial number of the SIM).

Next command group is about **user device status and control**, e.g., perform a reset, set power modes (incl. "airplane mode"), indicator for network or battery, real-time clock, boot behavior, SIM management.

Network service commands are dealing with network detection, selection, and configuration incl. eDRX and paging window settings, registration procedure, radio connection (RRC) status, received network signal strength, optimizations for network attach, TAU requests and coverage enhancement level, etc. A group of dedicated commands are provided for **packet-switched services**, i.e., the "PS domain" like GPRS for data transmission. A group of standard commands for SMS message handling is also available.

On top of commands dealing with basic function, manufacturers have added **proprietary AT command extension** to control various module features, e.g., for

- Firmware update
- Clock and power management
- GPIOs and other interfaces, e.g., I²C and for GPS engine
- Module file system administration
- IP sockets management
- Device and data security
- Data transmission protocols and message handling for SMS, FTP, HTTP, TCP, MQTT, CoAP
- Cloud services incl. LwM2M device management

For IoT application developers, the module's AT command set is the most relevant programming interface (API) because it allows to verify some NB-IoT network settings and to control some of them. Usually, application developers do not have direct access to standardized low-level procedures, e.g., for network resource allocation, assignment of coverage enhancement level, or relevant parameters for power reduction (see section "NB-IoT Technology" of chapter "Cellular IoT Technology"). Instead, each module offers an exclusive set of functions which are **NB-IoT application toolkit** at the same time. Besides other criteria mentioned in Fig. 3 of section "NB-IoT Cellular Network Modules," the module's unique AT command set is an important short-listing aspect.

For practical evaluation please also refer to section "RasPi Mockup and Network Tester" of chapter "Designing an NB-IoT Device" introducing a tool which can be used for functional verification of standards AT commands. This tool is based on Quectel BG96.

Unsolicited Result Code (URC)

URC is a **modem message** which is not responding to an AT command which has been submitted by the IoT application before. Instead, it is kind of a soft interrupt indicating an unscheduled event or status change, e.g., an incoming SMS or other an IP data packet. Another potential event is, for example, start of Active Timer T3324 (see section "Power Saving Mode (PSM)" of chapter "Cellular IoT Technology") when PSM function was enabled and an RRC connection release has been received.

A URC can occur at any time to inform the application about a specific event or status change. Because completely uncorrelated to an AT command execution, a collision between a URC and an AT command might occur. Some modules allow users to configure URC feature, which event is causing a URC and how/when to present an occurrence. For example, an enabled URC can be buffered by the module until an AT command execution ends when the final result code for the command has been returned.

In addition to the URC message, configured events can also trigger an external signal. For this purpose, most network modules have a dedicated **output pin RI** (Ring Indicator). This is a powerful feature for some IoT devices and can be used, for example, as an interrupt signal to wake up the host MCU.

IoT Data Transfer Protocols

Main reason for IoT connectivity is data transfer from a remote location to a central server. While the existing structure of the Internet is freely available and usable by any IoT device, these instruments often too heavy and too power-consuming, esp. for battery-powered devices. Good old SMS messages are still good enough as a starting point, but is not available in all NB-IoT networks. Plain TCP/IP low-level

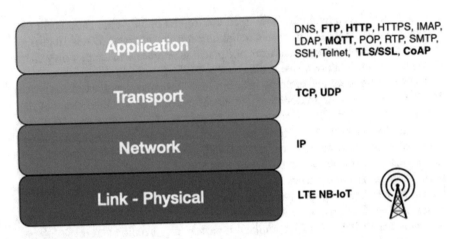

Fig. 6 C-IoT Internet protocol suite

data transmission allows efficient communication for small IoT deployments, but some IoT projects will require scalable solutions for a larger number of connected devices. First candidate is MQTT (Message Queuing Telemetry Transport) which is aiming data collection and deployment from/to IoT devices. Another option is CoAP (Constrained Application Protocol) which is aiming at a HTTP-type web presentation of IoT data. Both are application resp. service layer protocols which are operating on top of TCP/IP or another transport layer protocol (see Fig. 6).

IP-Based Protocols

Sending and receiving IoT data through the Internet sometimes require the cellular IoT device to behave like an IP network node, i.e., as an Internet endpoint with an IP address (IP = Internet Protocol). For this purpose, most NB-IoT module vendors are offering an integrated protocol stack which is a set of services that allow processes to communicate over the Internet using the protocols (e.g., TCP, UDP) offered by the stack. The module OS forwards the payload of incoming IP packets to the corresponding IoT application by extracting the socket address information from the IP and transport protocol headers and stripping the headers from the application data. For cellular network modules, used API for interaction with the integrated IP protocol stack is part of the AT command set, i.e., a proprietary extension of the AT command set.

Typically, an IP endpoint is implemented as a "socket," there are several types of IP sockets: Datagram sockets, Stream sockets, and Raw sockets. Stream sockets are connection-based and provide a sequenced flow of error-free data packets, reliably and in right order. **Stream sockets** are typically implemented using TCP (Transmission Control Protocol) so that applications can run across any networks using TCP/IP protocol. **Datagram sockets** are connectionless with data packets sent and received individually addressed and routed one-by-one. User Datagram Protocol (UDP) is used. Order and reliability are not guaranteed with datagram sockets, so multiple packets sent from one machine or process to another may arrive in any order or might not arrive at all.

UDP is a light-weight protocol offering high throughput and low latency (vs. TCP) which is good for unidirectional broadcasting of audio or video files, but cannot be used for a reliable IoT communication channel. On the other side, TCP provides end-to-end reliable communication incl. correction of transmission errors based on error detecting code and an automatic repeat request (ARQ) protocol. On top of this, TCP manages reordering of data packages which have received out-of-order. As a consequence, TCP is used for many popular applications, including HTTP web browsing and email transfer.

For IoT applications, some low-level services for TCP and UDP transport layer protocols are available via AT command interface, e.g., sending and receiving raw data which is not wrapped in any upper layer overhead data. In order to prepare an IoT device on transport layer for TCP or UDP communication, we first have to set up the device as an IP network node and create an **IP socket**. Different modules will offer different configuration options, but in general this socket will be reachable by its IP address by any external TCP/IP node, and will be listening for incoming traffic.

Typical AT commands for TCP/IP communication are (syntax from Quectel):

- Open a Socket Service **AT+QIOPEN**
- Query Socket Service Status **AT+QISTATE**
- Send data **AT+QISEND**
- Retrieve the Received TCP/IP Data **AT+QIRD**
- Ping a Remote Server **AT+QPING**

If an open socket has been configured as for TCP or UPD listening and new data has been received, the IoT application will be notified by associated URC message and/or RI interrupt (see also Fig. 7). In order to avoid collision of the URC with the ongoing execution of a submitted AT command, incoming data can be buffered temporarily by the NB-IoT module until the IoT application is ready for reception.

For uplink data transmission, the NB-IoT device must connect to the IP address of the recipient node first. After successful connection, the IoT application can submit uplink data via AT command to the modem. The modem can be configured to acknowledge successful data submission to the cellular network via related URC message to the IoT application.

MQTT (Message Queuing Telemetry Transport)

MQTT is a widely adopted standard in the Industrial IoT, for meters and detection devices and vehicles. It is a **publication/subscription type** messaging protocol. MQTT has been designed for communication in low-bandwidth networks, it has a small code footprint and requires low processing power and memory, i.e., an MQTT

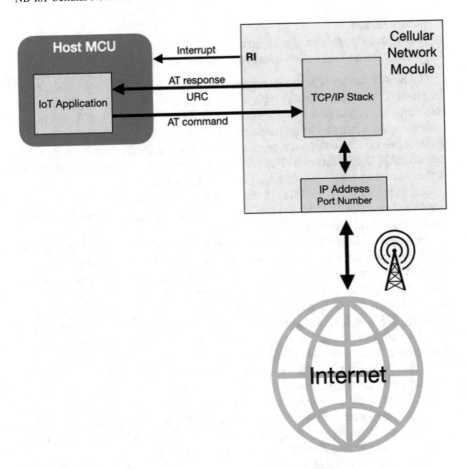

Fig. 7 TCP/IP for cellular IoT

client running on an IoT device will cause only low-power consumption. The protocol usually runs over TCP/IP; however, any network protocol that provides ordered, lossless, bi-directional connections can support MQTT. The protocol is an open OASIS standard and an ISO recommendation (ISO/IEC 20922).

The MQTT protocol defines two types of network entities: a message **broker** and a number of **clients**. An MQTT broker is a server that receives all messages from the clients and then routes the messages to the appropriate destination clients. Information is organized in a hierarchy of **topics**. Whenever one of the clients has a new item of data to distribute, it sends a **control message** to the connected broker including the new data and associated topic. The broker then distributes the information to any clients that have subscribed to that topic.

A MQTT control message can be as little as two bytes of data, but can also carry nearly 256 megabytes of data if needed. There are 14 defined message types used

- To connect and disconnect a client from a broker,

- To publish data,
- To acknowledge receipt of data, and,
- To supervise the connection between client and server.

The MQTT broker is software running on a computer, either on-premises or in the cloud. Clients only interact with a broker, not with other clients. The **broker role** needs to be specified by an **agreed policy**, but in general it will be up to the broker to ensure integrity of all participating clients and to ensure security and reliability of service. In addition, a certain level of **service quality** can be part of the defined MQTT infrastructure. For example, the broker has to ensure that a subscriber receives a message only once (i.e., no duplicates).

Connected MQTT clients do not need to know each other and do not communicate directly with each other. Instead, topics are used to categorize messages, authorized clients can subscribe to.

For a typical Remote Monitoring/Detection use case scenario, a sensing IoT device will publish local measurement data to the MQTT broker. For example, device X monitors the actual ambient temperature of 21 °C in room 5 on floor 1 in office building B. In order to classify this information, data will be published by device X as a hierarchical topic containing several levels like **/temperature/ building_A/Floor_1/Room_A105**. This is a multi-level topic, clients can subscribe to each level, whereas upper levels include subscription to respective sublevels, e.g., a subscriber of **/temperature/building_A/Floor_1** will receive temperature data for all rooms on floor 1. As the tenant T of room A105, you might be interested in this particular data only, so—as a first step—you contact the responsible MQTT broker B and apply for connection. After successful verification of authorization and identity of tenant T, the smart phone of tenant T is allowed to subscribe to topic **/temperature/building_A/Floor_1/Room_A105**. Consequently, broker B will forward messages of device X to the smartphone of tenant T.

Situation looks different for property management P of building B who is interested to monitor temperature of all rooms in building B, so P can subscribe to **/ temperature/building_A** and will receive all temperature data messages of connected and publishing IoT sensors in building B. As an alternative, data might first go to a connected analytics and data consolidation service S (see "Analytics" icon in Fig. 8). For this purpose, S will subscribe to **/temperature/building_A**, perform agreed work and publish resulting data each month as a separate hierarchical topic, e.g., as **/temperature/building_A/July2021**.

From an IoT application point of view, many NB-IoT network modules are offering a set of MQTT-related AT commands allowing the device to manage data accordingly. Each vendor implements MQTT AT commands in a different way. Typical MQTT commands are with syntax from Quectel are as follows:

- Connect a Client to MQTT Server **AT+QMTCONN**
- Subscribe to Topics **AT+QMTSUB**
- Publish Messages **AT+QMTPUB**
- Read Messages from Buffers **AT+QMTRECV**

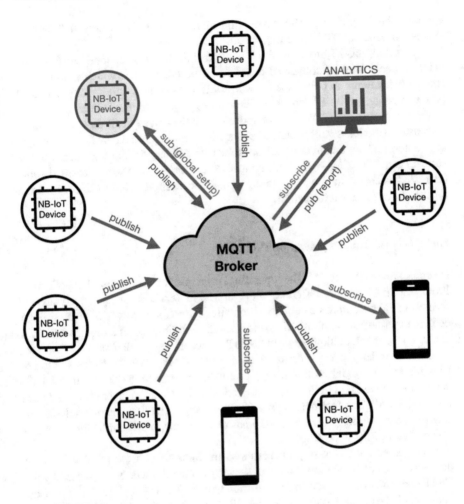

Fig. 8 MQTT roles

Similar to process for TCP/IP downlink data, new MQTT subscription data forwarded by the MQTT broker will be indicated to the IoT application by an associated URC message and/or RI interrupt (see Fig. 7).

Each client can both produce and receive data by both publishing and subscribing, i.e., an IoT device can publish sensor data and still be able to receive configuration information or control commands which are valid for all or a specific group of IoT devices. Although many NB-IoT applications are focusing on uplink data (see section "NB-IoT Use Cases" of chapter "IoT Target Applications") based on a preconfigured factory setup which is not altered during product lifetime, the MQTT option to change client's publish/subscribe roles will be useful for many NB-IoT applications—at least occasionally, e.g., for a changed PLMN list to be considered by all devices in the field (see case "global setup" in Fig. 8). Due to MQTT's flexible approach how to manage client roles and hierarchical topics, many IoT deployments can adjust MQTT according specific need.

In addition, MQTT broker model is beneficial for IoT applications which require **privacy protection** of connected clients. The broker decides which kind of data about other clients is being shared with other clients, and will specify a related policy to be agreed by all connected clients resp. owners of these clients.

On top of this, MQTT uses **Transport Layer Security (TLS) data encryption** with user name, password protected connections, and optional certificates for used key material. This architecture allows to set up IoT applications to meet highest security requirements—with extra tamper protection for critical use cases if some dedicated security hardware is being used. See section "End-to-End IoT Data Security" for further information.

CoAP (Constrained Application Protocol)

In comparison to MQTT, the operating principle of Constrained Application Protocol (CoAP) is very different: while an MQTT server (broker) is pushing subscribed IoT data updates to clients automatically, a CoAP server is waiting for a dedicated request by an (authorized) IoT client each time. Such a request will lead to a one-to-one interaction between the IoT device and an associated CoAP server.

CoAP was designed to address the needs of HTTP-based IoT systems, but translates the HTTP model so that it could be used in restrictive (aka "constrained") device and network environments. CoAP relies on the transport layer User Datagram Protocol (UDP). The Internet Engineering Task Force (IETF) Constrained RESTful Environments Working Group (CoRE) has done the major standardization work, it is specified in RFC 7252 (Fig. 9).

CoAP is essentially a **request/response protocol** to be used by an IoT client device with an existing CoAP Internet server. Connection will be requested by the CoAP client as a one-to-one communication with a specific CoAP server. If the host component is provided as an IP-literal or IPv4address, then the CoAP server can be reached at that IP address, e.g., via **coap://<IP address:port>**, resp. **coap://<host name:port>**.

A CoAP client can use the GET, PUT, POST, and DELETE methods using requests and responses with a CoAP server. Depending on use case (see sections "Object Tracking/Localization" resp. "Remote Monitoring/Detection" of chapter "IoT Target Applications"), an IoT client can either POST data or PUT data. The IoT device will POST it, if submitted IoT data is new and should be created by the CoAP server as a new record. On the other hand, a PUT request can be used to insert data resp. replace if it already exists. For example, a temperature sensor would PUT a periodic update rather than POST it. In general, each CoAP requests and response message may be marked as:

- "Confirmable" (CON): the messages must be acknowledged by the receiver if successfully received or as

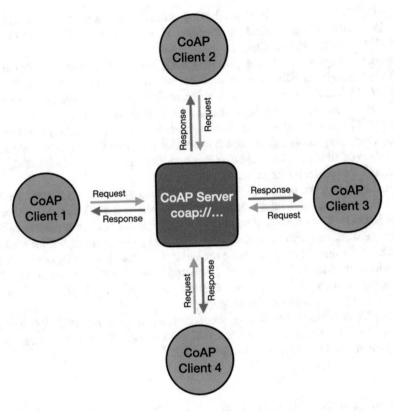

Fig. 9 CoAP request/response principle

- "Non-confirmable" (NON): the messages are "fire and forget."

This CoAP option allows IoT applications to meet requested quality of service resp. shorten interaction time and reduce corresponding power consumption of requesting client devices when using NON messages.

Different network module vendors are implementing their CoAP client APIs differently. For u-blox SARA-N3 module [17], for example, client profiles are used to define relevant parameters for each interaction with a specific server. Up to four profiles can be stored in the module flash memory and only one can be loaded at a time. The loaded profile will be considered as the current profile to be used for subsequent CoAP client requests. For profile management, **CoAP Profile Configuration Command +UCOAP** is used. Syntax: **AT+UCOAP=<op_ code>,<param_ val>** is used. For op_code=1, the URI (Uniform Resource Identifier) for this profile is specified. For example, this command will set the destination path of a CoAP profile to a resource called "text." **AT+UCOAP=1,"coaps://1.123.123.123:5684/text"**.

Now, **CoAP command +UCOAPC** can be used send a client request to the CoAP server. Syntax: **AT+UCOAPC= <coap_command>.** Allowed values for <coap_command> are "1" for GET request, "2" for DELETE, "3" for PUT, and "4" for "POST." So, for example, the following command will write "Hello World" (in ASCII format) to the destination specified in above profile: **AT+UCOAPC=3, "48656C6C6F20576F726C64,",0**.

Another op_code ("8") option of CoAP Profile Configuration Command +UCOAP enables an **SSL connection** between CoAP client and server. For example, **AT+UCOAP=8,0** specifies to use Security Profile 0 specifying all relevant TLS/SSL parameters. It will be activated whenever the **coaps://** scheme is used.

Note: TLS communication protocol is based on cryptographic keys and certificates which are potentially vulnerable to attacks. For critical IoT applications, extra protection of sensitive data might be required. See section "End-to-End IoT Data Security" for further information.

Last but not least, op_code ("9") option of CoAP +UCOAP command enables Release Assistance indication (RAI) feature of NB-IoT. For example, command **AT+UCOAP=9,1** will set RAI flag to "1" instructing the modem to manage release of the network connection to RRC_Idle state and switch off radio right after the uplink data is sent. This makes sense for battery-powered monitoring devices sending infrequent snapshot measurement data and go back to low-power mode afterwards. See section "Release Assistance Indication (RAI)" of chapter "Cellular IoT Technology" for background information.

Note: By nature, this option cannot be selected with confirmable (CON) message type because in this case the modem will have to wait for acknowledgement by CoAP server and has to keep radio switched on.

End-to-End IoT Data Security

In order to ensure bullet-proof communication and remote control of IoT devices, IoT projects will have to implement an appropriate level of security and protection against potential attacks or attempts to misuse the IoT application. Independent from technology used for data transmission, a secure channel between communication partners will have to protect integrity and confidentiality of data. This end-to-end security will ensure that nobody can understand or modify data transferred from one endpoint to another (typically from an IoT device to a dedicated IoT server or vice versa). Typically, an IoT server is located in a safe environment. But for many IoT use cases, end-to-end security requirement is particularly challenging because involved IoT devices are mobile and/or unattended, i.e., exposed to risk.

For use in IoT devices, all vendors of cellular network modules are including features or options to support TLS (Transport Layer Security), the successor of SSL (Secure Sockets Layer), see Fig. 10. TLS is a secure communication protocol which is using **public-key cryptography.** For an NB-IoT device, in most cases the endpoint of a secure TLS channel will be an integrated "secure element" inside the network module. For a software-only implementation, the TLS/SSL software stack will be part of the module firmware and accompanied with a dedicated set of AT

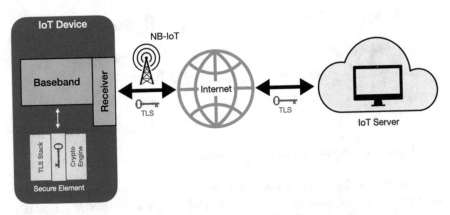

Fig. 10 Secure element for TLS-secured channel

commands for use by the IoT application. Key storage and crypto operations will be performed by the module MCU.

How Public-Key Cryptography Can Help

A fundamental ingredient of secure communication systems is **public-key cryptography** (or asymmetric cryptography) which is an encryption scheme. Other than symmetric cryptography, public-key cryptography uses a **pair of keys** which are different but mathematically linked to each other: i.e., a *public key*, which may be disseminated widely, and a *private key*, which are known only to the "owner." In a one-to-one electronic communication one of the keys (either public **or** private— depending on use case) is used by the sender, the other one is used by the recipient. Difference is illustrated in Fig. 11.

Public-key cryptography has a big advantage when compared to classic symmetric crypto schemes where same unique secret key is used by sender as well as by recipient: effective security requires keeping only the private key as a secret; the public key can be openly distributed without compromising security → key distribution and key storage are much easier to handle. But due to the fact that one key is public, asymmetric crypto algorithms are more complex and require longer key lengths compared to symmetric cryptography— at the same level of security. For example, security provided with a 1024-bit key using asymmetric RSA is considered approximately equal to an 80-bit key in a symmetric algorithm like AES.

In general, public-key cryptography can solve different fundamental security problems related to one-to-one communication scenarios:

1. Protect message **privacy**, i.e., nobody else is able read contents of message
2. Verify **authenticity** of sender, i.e., sender identity is tamper-proof and unique
3. Ensure **integrity** of message, i.e., make sure that nobody is able to modify message contents on its way to the recipient

1) symmetric

A ⟷ B

secret   secret

same key

2) asymmetric

A ⟷ B

public   private
secret

key pair, different keys

Fig. 11 Keys for symmetric vs. asymmetric encryption

Problem 1 can be solved by **data encryption**, i.e., sender uses the public key of the recipient → decryption of the received message can be done with the recipient's private key only. Problems 2 and 3 can be solved by use of a **digital signature**—to be applied using the sender's private key.

Message Authenticity and Integrity

In general, the sender's private key is used to sign a message, and all recipients can verify the sender's authenticity with the sender's public key. In order to limit required computational effort to create a digital signature, only the **hash** value of the message is encrypted, not the complete message itself. What does "hash" mean? A hashing algorithm is a mathematical function that condenses data to a fixed size, a hash is a fingerprint of the original data. On top of that a **secure hash** is irreversible and unique. Irreversible means "one-way," i.e., from the hash itself you could not figure out what the original piece of data was, therefore allowing the original data to remain secure and unknown. Unique means that two different pieces of data can never result in the same hash value. Today, for digital signatures SHA-2 algorithms are common, e.g., SHA-256 with a hash length of 256 bit.

So, **on sender side** (see Fig. 12) the message will be hashed, encrypted with sender's private key (aka "signed") and attached to the message before being transmitted via a public (i.e., unsecured) network. In order to verify sender authenticity and message integrity this signature will be decrypted with the sender's public key **on receiver side**. This operation creates H*(M). For verification purposes the received message M* will be hashed using the same hash algorithm as on sender side. This auxiliary hash H(M*) must be identical to the received decrypted hash H*(M). If not equal, very obviously something is wrong, either because

1. Used public key on sender side does not match → identity of sender is questionable

   or

2. Received message is not identical with original message → message content is questionable

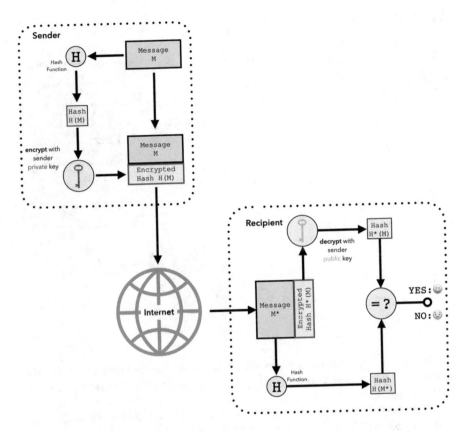

Fig. 12 Digital signature—create and verify process

Public Key Infrastructure (PKI)

In general, a public key represents the **electronic identity** of a person or an object. With a public key you can verify a digital signature, i.e., authenticity of the sender, but the key itself does not provide any information about the sender. This is why a **public key infrastructure (PKI)** is required for every application using public-key cryptography.

A PKI consists of a set of policies and an associated system that manage the creation, distribution, revocation, and administration of electronic identities incl. corresponding key pairs. PKI is about how two entities can trust each other in order to exchange messages securely. Usually, this is done by means of delegated trust using **certificates** issued be a mutually trusted entity, a so-called **certification authority (CA)**. Each certificate links a public key to the corresponding electronic identity, it contains relevant information required for the application to work properly and in a trustworthy and tamper-proof manner (Fig. 13).

A PKI has to specify how new electronic identities (i.e., for new IoT devices) are added to the application environment and how to revoke obsolete or expired

Fig. 13 PKI certificate

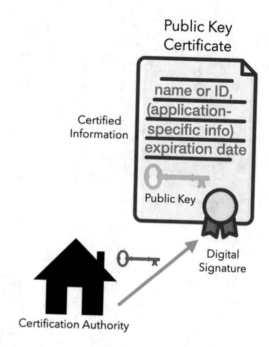

certificates. In any case, an up-to-date certificate database must ensure that all valid public keys are available to all participants, i.e., all potential message recipients, e.g., via central online repository.

For large-scale roll-outs of complex applications like a national citizen ID card a suitable PKI will be very complex, but for many IoT applications requirements might be much more simple and easier to implement, e.g., you might only one central registration authority which is exclusively handling certificates for new IoT devices and all IoT devices in the field. A PKI for a specific IoT project might be tailored to application requirements and should be as simple as possible, but at least you will have to consider, specify, and implement all relevant aspects how to handle and deploy public keys for all communication endpoints within your application environment.

TLS Handshake and DH Key Agreement

TLS key ingredients are cryptographic algorithms (asymmetric, symmetric, hash) and a PKI infrastructure for public key management and deployment. Symmetric crypto is used for data encryption because used keys are much shorter and reduce required computational effort which is particularly beneficial for low-cost and battery-powered IoT devices. Because with symmetric cryptography both parties have to use the **same crypto key** for data encryption as well as for data decryption, a major challenge is to agree or to exchange a session key securely. Traditionally, this was done by physical means, e.g., by a trusted courier delivering key lists written down on paper. The Diffie–Hellman key exchange method allows two parties that do not know each other to jointly establish a shared secret key over an insecure channel like the Internet.

For preparation of a TLS/SSL-based secure communication session, IoT client and server have to perform a handshake protocol to exchange parameters and shared keys. Key parameter is the **cipher suite** to be used for exchanging messages. A cipher suite specifies used cryptographic algorithms and key lengths. For example, TLS_DHE_RSA_WITH_AES_256_CBC_SHA means:

- Tunnel type: TLS
- Public-key algorithm for digital signatures and PKI: RSA
- Key exchange method: DHE (Ephemeral Diffie–Hellman)
- Symmetric algorithm for data encryption: 256-bit AES with CBC
- Hashing method: SHA

As a prerequisite before starting the TLS handshake, both parties have exchanged and mutually verified validity of provided certificates for used public keys of IoT client and IoT server. This means that both parties have authenticated each other, so they are prepared for handshake between trusted communication partners. Messages will be readable by anybody "in the middle," but they are digitally signed, so the message content cannot get manipulated.

In Fig. 14 a sample **Diffie–Hellmann (DH) handshake** for a secret key agreement is illustrated. Circled numbers in picture are referring to numbers in brackets in text. For a sample calculation of an agreed pre-master secret, parameters with the following values have been used:

- Modulus **p** = 25 (N)
- Base **g** = 38 (G)
- Client: secret random **a** = 2
- Server: secret random **b** = 5

After mutual authentication with the server has been done, the IoT client device will start the TLS handshake process by sending a list of cipher suites supported (1) to the server. In response, the server will return a message indicating which cipher suite has been selected for secure messaging (1).

In this case we are using a handshaking method of DHE-RSA, a 256-bit AES-CBC shared key, and with a SHA hash signature. In order to generate a symmetric session 256-bit AES key for data encryption, the protocol uses the multiplicative group of integers modulo **p** (aka "modulus") where p is prime, and **g** is a primitive root of prime number p (aka "base"). Numbers g and p are random, but carefully selected as seed parameters for the calculation process and will be shared with the client (2).

The server will then generate a random number **b**, and based on previously generated values for g and p, the server will then generate $B = g^b \bmod p$. On client side, same operation is done with secret random a: $A = g^a \bmod p$. Both results A and B are shared with the other party (3).

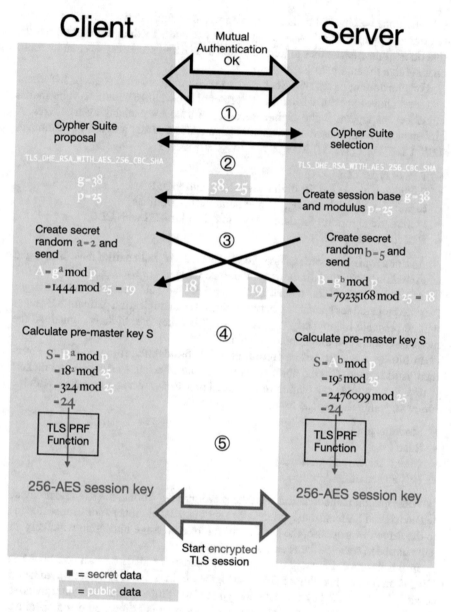

Fig. 14 DH key agreement (handshake concept)

Now, the trick is that on both sides the same secret S = g^{ab} **mod p** can be calculated:

$S = A^b \bmod p = (g^a \bmod p)^b = g^{ab} \bmod p$
$S = B^a \bmod p = (g^b \bmod p)^a = g^{ba} \bmod p$

Only a and b are secret, all other values are sent in clear. $S = g^{ab} \bmod p$ is called the pre-master secret (4).

Here, for demonstration purposes, in this example only small numbers have been used. With $p = 25$ only 25 possible results of a pre-master secret are possible ($n \bmod 25$) and can be determined shortly. However, if p is a large prime of—let us say—1000 bits, then even the fastest known algorithm on fastest computer cannot find random number **a** based on given public values of g, p and ($g^a \bmod p$) and ($g^b \bmod p$).

Based on this shared pre-master secret, both client and server can generate a master key. For this purpose, the TLS PRF (Pseudorandom Function) will be used. In TLS 1.2 this is created using an HMCA-SHA256 hashed value which will generate a 256-bit key. To create the actual key used we feed the master key and the nonce into the PRF and generate the shared 256-bit AES key for the session (5).

For use of the module TLS/SSL stack, if available, a proprietary AT command extension is provided as part of the firmware. As an example, for Telit ME910G1 NB-IoT module [18] the following commands are available to communicate securely with a remote SSL server:

- Configure Security Parameters of an SSL Socket **AT#SSLSECCFG**

  - Select cipher suite and authentication mode (server and or client).

- Manage the Security Data **AT#SSLSECDATA**

  - Stores, reads, and deletes security data (certificates, private keys) in/from module NVM (non-volatile memory)

- Enable an SSL Socket **AT#SSLEN**
- Open an SSL Socket to a Remote Server **AT#SSLD**
- Send Data through a SSL Socket **AT#SSLSEND**
- Read Data from an SSL Socket **AT#SSLRECV**

As a default setting, TLS is focusing on authentication of the IoT server, i.e., the client verifies the identity of the server. This might be fine for traditional browser–server interaction, but for IoT applications also user devices are potential candidates for hacking and identity theft. As a consequence, mutual authentication is required to cover IoT devices as well and extend overall security scope. **Mutual TLS (mTLS)** is an option of the TLS protocol which can be used for this purpose. With mTLS, a two-way authentication of both parties is performed at the same time (using a challenge-response approach) and must succeed before any data exchange can start.

Security Hardware and Certifications

For most TLS stacks the module MCU will be used for cryptographic work and module memory will be used for key storage (see Fig. 10 in section "End-to-End IoT Data Security"). Even if no extra protection for cryptographic material and processes has been implemented, software solutions are good enough for many IoT use cases. But in general, standard MCUs and memory products are not designed to provide high-level tamper protection against well-educated attackers with expensive equipment. Faking device identities and IoT messages are potential threats, but sometimes it is just "brand protection" why manufacturers want to prevent people from cloning an IoT device.

Some IoT projects are at high risk, for example, because they trigger flow of money. Misuse or tampering involved IoT devices might cause financial damage. A popular IoT use case is smart metering, i.e., remote access to household electricity meters where transmitted consumption data is automatically converted into an energy bill which is addressed to the registered customer— without any human interaction or control. Needless to say, that energy providers are interested in an efficient infrastructure, but incorrect data delivery would cause financial loss resp. legal/liability problems. As such, this kind of IoT application requires strong protection.

But what does "strong protection" mean? On the market you will find some MCUs and MCU subsystems selling their products as "secure elements" emphasizing implemented security and crypto performance for use of their products as a "trusted zone" incl. protected storage and a secure operating system. For further evidence some manufacturers of secure elements are offering security evaluation results provided by independent crypto experts, e.g., a CC certificate (CC = "Common Criteria for Information Technology Security Evaluation," short: "Common Criteria" or "CC").

Common Criteria is an international standard (ISO/IEC 15408) for computer security certification [19]. Common Criteria is a process how to specify security functional and assurance requirements for IT products. A user (e.g., a governmental institute or organization) can formally specify security requirements as an implementation-independent Protection Profile (PP). Each PP is addressing a specific use case scenario or application, e.g., a digital tachograph for vehicles, a machine-readable travel document (ePassport), or a cash register. A PP may cover hard- and software components and contains threats, security objectives, assumptions, functional requirements, and security assurance requirements. For compliant products manufacturers will have to implement these requirements and submit candidates for certification to accredited CC testing laboratories for final verification. Specified CC Evaluation Assurance Level (EAL1 through EAL7) reflects how thoroughly products must be tested, i.e., the quality of implementation. Of course, the CC certificate will not disclose any implementation details, but—in combination with applied Protection Profile–it will tell which security measures have been taken, e.g., which kind of attacks have been addressed, e.g., physical intrusion, fault attacks, side-channel attacks, power analysis (see [20] for further explanation).

In fact, a CC evaluation is common practice for many national smartcard projects used for identification purposes. For this kind of projects, a successful security evaluation is mandatory for suppliers who want to qualify their products for related tenders. This business scenario mainly applies to governmental smartcard roll-outs, but in the meantime, CC security evidence is also required for some IoT projects. For example, the German nation-wide electricity meter roll-out is asking for a level EAL4+ **Common Criteria security certificate** for a so-called Smart Meter

Gateway (SMGW) to be deployed in households and industrial sites. An embedded smartcard is used as a security module for crypto operations and secure storage, this smartcard module will have to be certified even on CC level EAL4+. See [21] for more information about applied Protection Profiles (PP) for German Smart Meter Gateway IoT project.

In fact, many initiatives all over the world have started to regulate security of IoT devices. For example, the IoT Cybersecurity Improvement Act of 2020 (IoT CIA) is on the way in the USA since Dec, 4, 2020 and relies on the National Institute of Standards and Technology (NIST) to formalize security requirements for IoT devices which are owned or controlled by US government. This is another indication for increasing importance and market relevance of **proven IoT security**. Some manufacturers of cellular IoT network modules have started to prepare their products accordingly and provide evidence about strength and quality of implemented protection measures. For example, u-blox SARA-R5 (see Fig. 3) is featuring an embedded discrete secure element which is certified CC EAL5+. This extra chip is offering data protection, anti-cloning and secure boot. On top of this, it is acting as a "root of trust" for cloud services, e.g., secure firmware update (see section "Server Hosting and IoT Clouds"). Nordic Semiconductor nRF9160 does not offer any security certificate, but is based on an ARM Cortex-M33 MCU core and "CryptoCell" technology which is offering extra cryptographic and security resources for energy-constrained devices. ARM cores are used by many smartcard ICs and CC-certified smartcard products.

In general, developers can add an extra 1-EUR IoT Secure Element ("SE") to any IoT device design—independent from selected network module. All of these chips are CC EAL4 or EAL5 certified. SE chips will store credentials (e.g., private RSA keys for TLS communication) and run crypto algorithms securely and efficiently in a protected on-chip environment, sensitive data will never leave. Secure elements connect internally to the network module or to the host MCU via I²C or SPI bus. SE chips consume only few microamps in power-down mode and should wake up in active IoT periods only. This means that also battery-powered IoT devices can benefit (see "Estimate overall power consumption" and two design concepts). CC-certified secure elements and associated services, e.g., for device provisioning, are available from

- NXP Semiconductor, URL: https://www.nxp.com/products/security-and-authentication
- STMicroelectronics, URL: https://www.st.com/en/secure-mcus/authentication.html
- Infineon, URL: https://www.infineon.com/cms/de/product/security-smart-card-solutions/optiga-embedded-security-solutions/optiga-trust/

In fact, secure elements offer ultimate end-to-end security for IoT communication channels (e.g., on top of TLS) and a root-of-trust for IoT applications—at some extra cost. This investment might pay off for IoT devices requiring strong tamper protection.

Server Hosting and IoT Clouds

By nature, IoT applications are connected to the Internet, exchanging data between IoT clients and an IoT server. This is why an IoT project can take advantage of external online IoT services instead of implementing them in-house, e.g., server hosting, IoT device management, data analytics. Following increasing worldwide demand for IoT solutions, this market is versatile and encouraged many big players like Google or AWS to enter. Vendors of cellular network modules (modems) are bundling IoT services with their products in order to offer a one-stop-shopping experience to their customers.

An IoT cloud offers resources (servers, storage) and services to support operation of IoT applications and devices (Fig. 15). In general, IoT cloud services are leveraging available external expertise of IT companies and offload in-house development efforts to build an infrastructure for IoT device management and data processing. In fact, all IoT devices are delivering application-specific local data via Internet which require analysis and consolidation in order to generate actionable insights. This objective is particularly difficult to manage for large-scale IoT deployments with many devices and big data load. IoT cloud services allow IoT applications to collect, filter, transform, visualize, and act upon device data according to customer-defined rules. Another challenge is to manage deployed devices in the field, i.e., to check status, update functionality, or to re-configure them, if required.

By nature, IoT clouds offered by Google, AWS, etc. are generic, i.e., offered services are offered independently from target application and device hardware or used network technology. Services do not need any lower level support from device MCU, operating system, network interface, etc. Instead, on device side a TLS/SSL

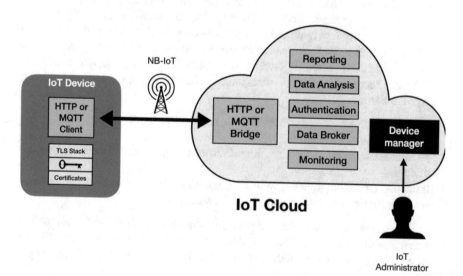

Fig. 15 IoT Cloud interface

socket is acting as a device identifier and endpoint for IoT cloud one-to-one communication. For Google and AWS IoT clouds (and others), HTTP or MQTT application layer protocols can be used for uplink IoT data transfer or downlink device updates. Embedded HTTP and/or MQTT clients are standard firmware functions offered by most NB-IoT network modules. Device identities can be securely stored and managed in combination with an integrated secure element and associated PKI certificates, if available (see sections "Public Key Infrastructure (PKI)" and "Security Hardware and Certifications").

A cloud function called "device manager" establishes individual device identities and authenticates the device when connecting. It also maintains a logical configuration of each device and can be used to remotely control the device from the cloud. Since the IoT cloud does not know any technical details of the NB-IoT device, each configuration request is must be converted locally by the host MCU into a sequence of module-specific AT commands.

IoT clouds are offered by

- Google Cloud IoT Core, URL: https://cloud.google.com/iot-core
- AWS IoT Core, URL: https://aws.amazon.com/iot-core/
- Microsoft Azure IoT, URL: https://azure.microsoft.com/en-us/overview/iot/
- IBM Watson IoT, URL: https://www.ibm.com/cloud/watson-iot-platform
- Telekom Cloud of Things, URL: https://iot.telekom.com/en/solutions/platform

and many others. In particular, vendors of cellular network modules like Telit or u-blox (see section "Vendor Overview") are predestinated partners for cloud-based services as an added value of their products. For example, u-blox SARA-R5 NB-IoT module has an integrated MQTT interface to AWS IoT cloud with its own AT command set for control by the device IoT application.

LwM2M Device Management

An interesting alternative is Lightweight M2M (LwM2M) which is a standard protocol from the Open Mobile Alliance (OMA) for **IoT device management and service enablement**. The LwM2M standard defines the application layer communication protocol between a LwM2M server and a LwM2M client which is offered by most cellular network modules. It offers an approach for managing IoT devices and allows devices and systems from different vendors to co-exist in an IoT-ecosystem.

Many industry members incl. module manufacturers and IoT platform vendors are OMA members, including ARM, Gemalto, Microsoft (Azure), Sierra Wireless, Telit, and u-blox. LwM2M's device management capabilities include remote provisioning of security credentials, firmware updates, connectivity management (e.g., for cellular), remote device diagnostics, and troubleshooting. LwM2M's service enablement capabilities include sensor and meter readings, remote actuation, and configuration of host devices.

Open source implementations of the LwM2M protocol incl. LwM2M server are available on GitHub. For more information see https://omaspecworks.org/what-is-oma-specworks/iot/lightweight-m2m-lwm2m/.

SIM Card or Embedded eSIM

Everybody knows that a new mobile phone will not work if no SIM card has been inserted before. In fact, every cellular device is identified on the mobile network by the subscriber identity stored in its SIM card (SIM = Subscriber Identification Module). A SIM card is a personalized electronic component which is associated with the selected connectivity partner, i.e., a network operator (MNO). The SIM is used to identify the user (and associated subscription plan), and it activates a pre-configured, MNO-specific connection profile on the user device. It also defines which service has been booked. Based on this commercial agreement (subscription plan), the selected network partner defines which network services are available and delivers those services using a combination of their own and/or sub-contracted mobile networks.

For a cellular IoT device, a SIM card is usually inserted during production or installation and will stay there until its end of life. Historically, SIM cards could hold just one subscriber identity linked to a single service provider. One service provider can still provide access to several networks through roaming, but in order to switch service providers, the SIM card needed to be physically changed to a new SIM card with a different subscriber identity. But, for some IoT use cases, SIM replacement would be practically impossible or this process would cost too much.

On the other hand, for some IoT projects flexibility of MNO selection during the complete device lifecycle would make sense. For example, if a company wants to permanently deploy devices in markets where permanent roaming is prohibited by regulation, such as in China and Brazil. Another reason is operational cost: even if an MNO offers roaming in an area which is not covered by its own network, a local MNO might offer IoT connectivity at a lower price.

Swapping SIM cards is a logistical challenge esp. for large-scale international IoT projects where you might have different MNO preferences for different locations of installed IoT devices. For stationary IoT devices this means that you might have to differentiate production process depending on target location, i.e., you will have to maintain different device versions and different BOMs with different SIM cards, i.e., SIM cards from multiple MNOs need to be managed in the supply chain. Even more challenging are mobile IoT applications or devices which are supposed to work everywhere (see "Design Concept #2: Object Localizer").

From a technical point of view, a SIM card is just a copy-protected 32-64 kB storage device which is owned and managed exclusively by the MNO. But in order to increase flexibility and reduce cost, industry members have been pushing to "virtualize" the SIM card. Result is an **embedded SIM (eSIM)** or **eUICC** (embedded universal integrated circuit card) which converts the removable SIM card into a fixed

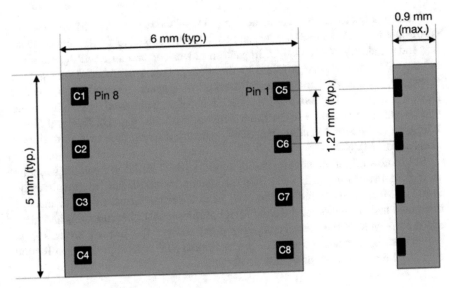

MFF2 bottom view

Fig. 16 eSIM MFF2 SMD package

SIM chip which can be soldered into the device PCB (Fig. 16). In fact, the eUICC is a single-chip secure MCU platform which is shared by MNOs in a way that multiple profiles can be loaded. Users can select and activate the most appropriate profile per remote control. The eUICC is a **secure element** based on a multi-application GlobalPlatform JavaCard operating system which is offering strictly separated security domains for each MNO who is owning and managing its content. eSIM standardization is being facilitated by GSM Association, further information and specifications are available here: https://www.gsma.com/esim/esim-specification/. An eSIM can accommodate **several subscriptions**, and provisioning of an eSIM can be performed over-the-air with clearly defined roles and interfaces described in GSMA's "Remote SIM Provisioning (RSP) Architecture for consumer Devices." High security standards for production and processes have been put in place in order to create confidence among all involved parties, i.e., chip manufacturers, device manufacturers, operators, service providers, and users. eUICC implementations are subject to Common Criteria evaluation aiming at assurance level EAL4+. Communication channels for **remote provisioning** are TLS-encrypted with authenticated parties based on PKI certificates with GSMA as the root certification authority.

The eSIM approach is addressing the patchwork and fragmented structure of the current global cellular network. For current IoT applications an eSIM will efficiently solve logistics problems in case a different MNO should be used. But eSIM might also create new ideas for new IoT business opportunities because it will allow IoT application owners to switch to the most appropriate network anywhere at any time— on the fly.

From practical NB-IoT device design point of view, use of eSIM will **reduce component count** and PCB space, and obsolete SIM connectors will increase security and reliability of the design. Impact on power consumption will be marginal because SIM card will be powered by the modem only when needed (see section "Design Concept #1: Environmental Sensor" of chapter "Designing an NB-IoT Device"). In general, massive eSIM adoption is ongoing and used by several mobile phones and also by some IoT cellular network modules, e.g., by Telit and Sierra Wireless (see section "Vendor Overview") who are offering their own subscription plans as IoT MVNOs.

But in general, implementing and operating eUICC-based IoT solutions require more than just dealing with a new SIM formfactor. Instead of just inserting a ready-to-use SIM card, fixed eSIM users will need a secure infrastructure to manage download and activation of different MNO profiles, subscriber identities and manage the lifecycle of the overall setup. For many straight-forward IoT projects, e.g., indoor and fixed-location use cases, the traditional SIM card approach is still good enough, if device space constraints do not prevail.

Sensors

Sensing one or multiple parameters of a remote location is a major ingredient of every IoT application (see section "NB-IoT Use Cases" of chapter "IoT Target Applications"). Typical parameters to be monitored are temperature, humidity, pressure, vibration, motion, fill level, weight, noise, volume, applied force, chemical composition, presence, distance, brightness, speed. But also, simple yes/no indicators might be relevant, e.g., an open door, presence of an object, full container. In fact, all NB-IoT network modules are offering various interface options for sensors (see Fig. 3). Typically, three different options are available: general-purpose I/Os (GPIOs), an integrated Analog/Digital Converter (ADC) and I2C or SPI serial data interfaces. A simple digital yes/no indicator can connect to a GPIO pin of the module. For analog signals, an ADC channel can be used. Digital sensors normally use an I2C or SPI interface for data transmission. A NB-IoT cellular module typically offer dedicated AT commands are usually available for all three inputs options. These AT commands are used by the IoT application software to control sensor operation and input data flow. If not supported by module AT commands, the host MCU will have to take over and provide suitable sensor interfaces.

For many IoT applications, IoT devices can deliver local sensor data on remote request at any time, but for battery-powered NB-IoT devices additional requirements apply in an effort to leverage PSM power saving feature. Limited battery capacity mandates a strict power reduced design with components remaining in power-down mode as long as possible. By default, the modem is in PSM mode and will wake up for scheduled or event-driven activity cycles only. Sensor devices should follow PSM cycles, i.e., enter stand-by operation whenever PSM mode has been activated by the IoT application software (in cooperation with NB-IoT

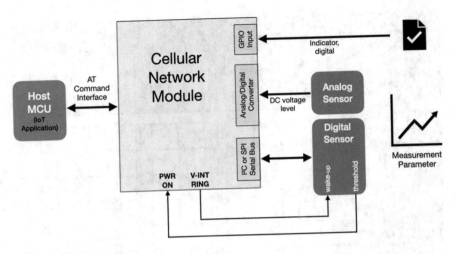

Fig. 17 Sensor interfaces

network). Independent from measured parameter, an IoT sensor should offer a power-down mode allowing the device to consume only few microamperes during this period. Activation can be done by software (I^2C command) or via dedicated hardware wake-up pin (see Fig. 17). For this purpose, many network modules offer an external signal which is indicating that the modem has switched its radio on, i.e., has returned to active mode. This mechanism can be used for scheduled IoT device activity, e.g., for monitoring IoT application (used for "Design Concept #1: Environmental Sensor"). As an alternate unscheduled wake-up approach, the modem's "RING" indicator (for a received message) can be used to initiate sensor full operation.

For sensor-driven operation of the IoT device, a wake-up pin is offered by most NB-IoT modules (called WAKE or PWR_ON or similar). Pulling this pin for a certain duration will end PSM mode and reactivate full modem operation. This pin can be used for a typical IoT monitoring applications (see section "Remote Monitoring/ Detection" of chapter "IoT Target Applications") where users are focused on a local parameter to escape from an expected range, e.g., ambient temperature gets too high. Thus, he/she wants to get alerted whenever a certain minimum or maximum value (threshold) is exceeded. Many digital sensors offer this function and the option to fire an interrupt signal in this case.

Lowest power consumption, wake-function, and threshold-driven interrupt capability are additional requirements which apply mainly for battery-powered zero-touch IoT applications. In general, important selection criteria for digital sensors are measurement performance like resolution, accuracy, response time, long-term drift, etc. Figure 18 is providing an overview of major vendors and typical categories for monolithic sensor chips with digital output (e.g., I^2C) which have been designed for mass-market cost-sensitive IoT applications.

Many sensors are directly matching the objective of an IoT application. For example, for remote monitoring of CO_2 pollution, specialized sensor ICs are

Manufacturer (Website URL)	Temperature	Humidity	Optical	Proximity, Time-of-Flight	Gas, Liquid Flow	Accelerometer	e-Compass	Gyroscope	Pressure (Barometer)	Microphone	Ultrasound	Voltage	Radar	Magnetic	CO2	Current	Gas, Liquid Flow
Analog Devices (www.analog.com)	x		x			x		x						x			
Bosch Sensortec (www.bosch-sensortec.com)	x	x				x	x	x	x								
Infineon (www.infineon.com)									x	x			x	x	x	x	
Maxim (www.maximintegrated.com)	x																
Microchip (www.microchip.com)	x														x	x	
Murata (www.murata.com)						x		x	x							x	
NXP (www.nxp.com)	x								x					x			
ON Semiconductor (www.onsemi.com)	x		x														
Renesas (www.renesas.com)	x	x		x	x												x
ROHM (www.rohm.com)	x		x			x	x		x					x		x	
ScioSense (www.sciosense.com)	x	x			x										x		x
Sensirion (www.sensirion.com)	x	x			x										x		x
Silicon Labs (www.silabs.com)	x	x	x											x			
STMicroelectronics (www.st.com)	x	x	x	x		x	x	x	x	x	x	x					
TE Connectivity (www.te.com)	x	x			x	x										x	x
Texas Instruments (www.ti.com)	x		x	x								x	x	x		x	
Vishay (www.vishay.com)			x														

Fig. 18 Digital sensors—overview

available to deliver the actual **CO$_2$ rate** in a digital format. Integration of these sensors via I^{2}C bus is straight-forward.

Other applications can use sensor outputs as an input parameter for calculations, e.g., to determine the distance to an object. For this purpose, the **time-of-flight (TOF)** of an emitted light pulse and reflected by the object. A sensor detects the returning signal, and the total travel time determines its distance to the object. For a seamless integration into an IoT device, sensor manufacturers are usually offering a comprehensive solution package incl. application notes, design kits, source code, etc. in order to accelerate customer design.

Another example are MEMS sensors (MEMS = micro-electro-mechanical systems) which are based on semiconductor technology to measure mechanical force, i.e., convert it into an electrical signal. In fact, **MEMS accelerometers** measure linear acceleration. But they can also be used for specific purposes such as inclination and vibration measurements which are needed for "Predictive Maintenance" of machines or equipment. With MEMS accelerometers you can also address special IoT use cases, e.g., detect an object which has been moved from its assigned location, or detect free-falling condition of an object. Sensor measurement data might not immediately answer the question, but measurement data can be used to feed calculation and creation of meaningful IoT data.

Suppliers and Online Support

Good news for IoT application developers is that the "Internet of Things" has been elected by all market players as the #1 top priority application for the IT electronics business. All contributors like semiconductor manufacturers, network operators, distributors, IT service providers, etc. are trying to benefit from promising IoT market outlook and take their share. For IoT device designers this means that they can expect to receive a decent level of support for their engineering work.

On top of this, most manufacturers of electronic components and subsystems have learned how to handle a large number of different customer projects via sales partners or online channels. In fact, most chips are offered as standard products accompanied by a **comprehensive set of documentation**, evaluation tools, and a design kit. Objective is to provide self-explanatory material which is supposed to answer most questions in order to minimize customer need for one-to-one support. All relevant product information should be **published online and downloadable** via manufacturer website. Usually, it also allows customers to order product samples, evaluation boards, design kits, etc. directly.

In addition, and whenever needed, an **authorized dealer (distributor)** will be the day-to-day business partner and entry point for all kind of customer requests. Traditional distributors are independent and work as a supply partner offering additional customer services incl. stock management, consultancy, and technical support. As an alternative, commercial customer requests can also go to **online distributors** like Mouser or Digi-Key who do not offer any additional support, but competitive prices.

IoT design engineers are particularly online-minded and might take decisions to select a component based on information which have been extracted from online sources. In general, manufacturer websites are most important **self-service repositories** for product information and a common starting point for application designers to prepare for competitive product comparisons. Besides technical information like datasheets and user manuals also white papers, presentations, and videos are available for download. Design kits should include drivers, sample source code, schematics, guidelines for PCB layout, etc. For components like cellular network modules or sensors which are specifically addressing devices for IoT applications, many manufacturers are offering **IoT-specific application notes and design tips** in order to support implementation and to speed up customer time-to-market. In particular, they should explain how to perform application-specific adjustments, e.g., which features have been implemented to save power consumption and/or how to configure a NB-IoT network cell according to application requirements.

On top of product information, manufacturers should offer **interactive support services**. A popular online support instrument is a virtual community where people with a particular common interest meet online and exchange information. For this purpose, manufacturers of electronic components offer a community platform with discussion boards for product-related topics. These are places where users can ask questions and share material with other community members. **Communities** are managed by a company moderator and supported by product experts, but key aspect for success are contributions from other users. Community members will have to register, but hide their professional identity from others, i.e., they can participate anonymously. By nature, all contributions are published and might help multiple visitors.

For non-public support requests, some manufacturers are offering the option to submit a **private support ticket**. Each case will be handled one-to-one by a company employee and will be escalated to a product expert, if required.

NB-IoT Network Deployment

After finalization of 3GPP Release 13, start of massive NB-IoT deployment was delayed by some network hardware issues and uncertainties regarding migration of NB-IoT into 5G standard. On top of this, NB-IoT was suffering from competitive offers by non-cellular LPWAN technologies Sigfox and LoRa. According to market research company ABI Research, in the early days most IoT activity took place in the APAC region which accounted for 40% of all global LPWAN connections in 2018 and for nearly 97% of all NB-IoT connections worldwide. In particular, China has been a critical market for the early adoption and growth of LoRa and NB-IoT.

LoRa and Sigfox are still strong competition, but according to a 2019 report of ABI Research, NB-IoT and LTE-M will capture over 60% of the 3.6 billion LPWA network connections by 2026.

In the meantime, NB-IoT networks are available in most countries of the world. More information about the status of national NB-IoT network deployment are consolidated by GSMA. National deployment maps (see: https://www.gsma.com/iot/mobile-iot-map) are providing a first indication if at least one NB-IoT network is available in a target country or not. At the time of writing, these countries are already covered (in alphabetical order): Argentina, Australia, Austria, Bangladesh, Belarus, Belgium, Brazil, Bulgaria, Canada, China, Columbia, Croatia, Czech Republic, Denmark, Estonia, Finland, France, Germany, Greece, Hong Kong, Hungary, India, Indonesia, Ireland, Italy, Japan, Kazakhstan, Latvia, Lithuania, Luxembourg, Malaysia, Malta, Mexico, Netherlands, New Zealand, Norway, Pakistan, Poland, Portugal, Qatar, Romania, Russia, Saudi Arabia, Serbia, Singapore, Slovakia, Slovenia, South Africa, South Korea, Spain, Sri Lanka, Sweden, Switzerland, Taiwan, Thailand, Netherlands, Turkey, UAE, Ukraine, United Kingdom, USA, Vietnam.

Another URL https://www.gsma.com/iot/mobile-iot-commercial-launches/ is listing MNO names which have deployed an NB-IoT network. At the time of writing, these companies are running national NB-IoT networks (in alphabetical order): 3, A1, AIS, APTG, AT&T, Altice, BASE (Telenet), China Mobile, China Telecom; China Unicom, Chunghwa Telecom, Deutsche Telekom, Dialog Axiata, DNA, DU, Elisa, Etisalat, FarEasTone, Grameenphone, HKT, Kcell, Korea Telecom, Kyvistar,

LGU+, LMT, Maxis, Magafon, Melita, Mobily, Mobitel, MTS, NOS, Orange, Ooredoo, Optus, Proximus, Reliance Jio, Rogers, SFR, SingTel, Smartone, Softbank, Starhub, STC, Sunrise, Swisscom, T-Mobile, Taiwan Mobile, TDC, TIM, Telefonica, Telenet, Telenor, Telia, Telkomsel, Telstra, True Corporation, Turkcell, Viettel, Verizon, Vodafone, Wind, XL Axiata.

In order to prepare an NB-IoT device for potential target locations, for designers it will be important to know which NB-IoT carrier frequencies are being used. In fact, 3GPP has defined an initial list of supported bands set of frequency bands in Release 13. Some additional bands have been defined in 3GPP Releases 14 and 15, see Fig. 1.

Note: NB-IoT works in half-duplex mode. This means that downlink uplink (UL) and downlink (DL) traffic are separated in frequency, so the CIoT device either receives or transmits, i.e., not simultaneously.

For NB-IoT project owners, GSMA information about network coverage on country level is a good for initial confirmation and which regional MNOs are involved. But regional coverage does not guarantee proper network access at all target locations for NB-IoT devices. Especially in rural areas, coverage maps with finer granularity or precise locations of cell towers are required. Hundred percent

NB-IoT Band	Device Uplink (UL)	Device Downlink (DL)	regional usage						
			Europe	Asia Pacific	North America	Latin America	Independent States	Sub-Saharan Africa	Middle East and North Africa
B1	1920 MHz – 1980 MHz	2110 MHz – 2170 MHz							
B2	1850 MHz – 1910 MHz	1930 MHz – 1990 MHz	x						
B3	1710 MHz – 1785 MHz	1805 MHz – 1880 MHz	x	x		x	x	x	
B5	824 MHz – 849 MHz	869 MHz – 894 MHz	x	x					
B8	880 MHz – 915 MHz	925 MHz – 960 MHz	x	x	x				
B12	699 MHz – 716 MHz	729 MHz – 746 MHz	x	x				x	x
B13	777 MHz – 787 MHz	746 MHz – 756 MHz					x		
B17	704 MHz – 716 MHz	734 MHz – 746 MHz			x				
B18	815 MHz – 830 MHz	860 MHz -875 MHz	x						
B19	830 MHz – 845 MHz	875 MHz – 890 MHz	x						
B20	832 MHz – 862 MHz	791 MHz -821 MHz	x	x				x	x
B26	814 MHz – 849 MHz	859 MHz – 894 MHz	x	x					x
B28	703 MHz – 748 MHz	758 MHz – 803 MHz	x	x					
B66	1710 MHz – 1780 MHz	2110 MHz – 2200 MHz	x						
added in 3GPP Release 14:									
B11	1427.9 MHz – 1447.9 MHz	1475.9 MHz – 1495.9 MHz							
B25	1850 MHz – 1915 MHz	1930 MHz -1995 MHz							
B31	452.5 MHz – 457.5 MHz	462.5 MHz – 467.5 MHz							
B70	1695 MHz – 1710 MHz	1995 MHz – 2020 MHz							
added in 3GPP Release 15:									
B4	1710 MHz – 1755 MHz	2110 MHz – 2155 MHz				x			
B14	788 MHz – 798 MHz	758 MHz – 768 MHz							
B71	663 MHz – 698 MHz	617 MHz – 783 MHz				x			
B72	451 MHz – 456 MHz	461 MHz – 466 MHz							
B73	450 MHz – 455 MHz	461 MHz – 465 MHz							
B74	1427 MHz – 1470 MHz	1475 MHz -1518 MHz							
B85	698 MHz – 716 MHz	728 MHz – 746 MHz							

Fig. 1 NB-IoT frequency bands

reliable information is available from MNOs only, sometimes they even publish it online. Figure 2 is illustrating a typical scenario with a planned device location marked in a network coverage map provided by a regional MNO.

Different colors indicate quality of local network coverage resp. coupling loss which has significant impact on required device output power and power consumption. In this case, it will be subject to further analysis, if coverage within this particular cell is good enough. Each MNO is exclusively driving local deployment of network infrastructure and technology updates, only network owners are able to provide a detailed NB-IoT deployment status and schedules. This is particularly important for IoT projects aiming at locations which are not "mainstream," i.e., an environmental sensor in a forest. Only MNOs will be able to inform precisely if a specific device location is within reach of one of their NB-IoT capable LTE towers or not.

In addition to "official" MNO information, a tool called "CellMapper" (www. cellmapper.net) can provide further insights. **CellMapper** is a crowd-sourced

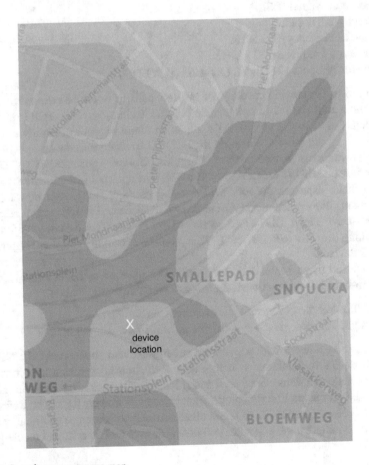

Fig. 2 Sample area coverage map

cellular tower and coverage mapping service. It works with Android smartphones and after download and activation, it runs in the background and collects connection data for upload. This works only for LTE networks, but since NB-IoT is "just" an LTE upgrade/extension, CellMapper might provide some useful information. Uploaded data will be used by CellMapper to extract details of individual connection with a network tower along with the position and technical details of that tower. Results are consolidated and displayed graphically on OpenStreetMap® geodata. Some relevant cell parameters are provided. This is, for example, data provided for a cell tower operated by MNO Telenor near Szentgotthárd in Hungary:

- eNB ID: 2690—LTE
- MCC/MNC/Region: 216/1/26131
- Cell Identifier: 688647
- System Subtype: LTE
- PCI: 226 (75/1)
- EARFCN: 6350
- Maximum Signal (RSRP): −98 dBm
- Direction: SW (205°)
- Uplink Frequency: 852 MHz
- Downlink Frequency: 811 MHz
- Frequency Band: EU Digital Dividend (B20 FDD)

On top of geographic coverage of NB-IoT cells, it might be relevant to understand the actual status of implemented NB-IoT technology resp. 3GPP release of each cell. For example, an NB-IoT application might expect availability of certain features, e.g., the NB-IoT Wake-up Signal (WUS) which introduced in 3GPP Rel. 15 or power class 3 (see section "Optimize Uplink Transmission" of chapter "Designing an NB-IoT Device") which is a Rel. 14 option. In addition, users should double-check expected network behavior and configurability, e.g., power saving mechanisms like immediate RRC release or max. accepted eDRX periods resp. values for Active Timer T3324 (see section "Low Power Device Design" of chapter "Designing an NB-IoT Device"). This applies especially for battery-powered NB-IoT devices.

Besides all this external data, for own local evaluation activities some commercial **handheld NB-IoT network meters** are available on the market:

- Exelonix *NB Meter* https://www.exelonix.com
- Keysight Technologies: *Nemo IoT Meter* https://www.keysight.com

Both products work with an Android smartphone and an inserted USB NB-IoT network device in combination with a dedicated Android app as a user interface. Both handheld solutions can be used at any location to check network coverage and determine parameters like Reference Signal Received Power (RSRP) or Reference Signal Received Quality (RSRQ) and other network performance and quality indicators. As an alternative, you can also build your own tool, see "RasPi mockup and network tester."

As already mentioned, national MNOs are key drivers of cellular technology in 3GPP and GSMA organizations and deployment of network infrastructure—resulting from increasing worldwide customer demand for NB-IoT solutions. Some **MNOs are actively promoting NB-IoT** technology with dedicated webpages and offers for application developers. Here are some of them:

- Deutsche Telekom: https://iot.telekom.com
- Vodafone: https://www.vodafone.com/business/iot/managed-iot-connectivity/nb-iot
- Telia: https://business.teliacompany.com/internet-of-things/iot-connectivity/LPWA-IoT
- China Telecom: https://www.chinatelecom-h.com

Depending on the geographic scope of an NB-IoT project, it might be difficult to select only one single MNO partner because network coverage of an alternate MNO might be better for certain locations. If, for example, in scenario illustrated in Fig. 2, a device would be operated by MNO 1 under CE level 2 conditions, but MNO 2 could be able to perform transmissions with CE level 0, so MNO 2 would be preferred. MNO 1 might be good for location 1, MNO 2 for location 2. For this scenario, it would make sense to work with both MNOs at the same time and have two subscription plans in place. National roaming would be more convenient, if available.

Roaming and MVNOs

In general, NB-IoT is predestinated for stationary devices rather than mobile devices. This means that a **local national MNO** partner is good enough for most NB-IoT deployments. But exceptions apply, for example, for use case addressed by our "Design Concept #2: Object Localizer." In this case, an NB-IoT device wakes up at an unpredictable location which is covered by a PLMN of a different country. In this case, international roaming will be needed for communication with this device. At time of writing, **NB-IoT international roaming** is offered by Deutsche Telekom and Vodafone covering all major countries in Europe, USA, Taiwan, and Australia.

For proper NB-IoT cross-border operation, MNOs will have to ensure that periodic Tracking Area Update (TAU) mechanism works fine with their roaming service. In addition, involved roaming partner network has to support PSM feature and has to maintain buffering of data packets sent to an NB-IoT device during PSM mode or eDRX periods (see section "Power Saving Methods" of chapter "Cellular IoT Technology"). Otherwise some downlink data might get lost when a recipient IoT device is in power-down mode resp. temporarily not listening to paging events.

Traditionally, cellular mobile network operators (MNOs) are offering services based on a network infrastructure they own and control exclusively. But the telco business model is moving and might end up with a mix of old and new players leveraging their particular expertise or end-user access. In the mobile phone business, **"virtual" network operators (MVNOs)** are offering cellular network services which are based on business agreements with network infrastructure owners. Depending on individual business objectives, each MVNO profile is different. In the meantime, also **specialized IoT MVNOs** have entered this promising marketplace with service packages which are tailored to the needs of IoT customers. One objective is cross-border network coverage for mobile IoT, e.g., for LTE-M networks. For example, EMnify has roaming agreements with a total of 700 networks in 180 countries, i.e., connectivity to several national MNOs is offered. This means that, for example, in France four different networks operated by four different MNOs are supported: Orange, SFR, Free Mobile, Bouygues. As a practical consequence and advantage, an IoT device at a certain location in France can select the most efficient IoT data connection out of four connectivity options.

On top of this extra flexibility of local operator selection, an **NB-IoT roaming** agreement allows IoT project owners to deal with one single operator (MVN or MVNO) with a **single SIM card** instead of multiple operators for different countries resp. for different device locations, each on them requiring an extra SIM card. This single-supplier approach more efficient and more convenient, as long as commercial conditions are competitive. For these advanced deployment scenarios, use of an eSIM is beneficial because **eSIM allows remote management of operator profiles on the fly** (see section "SIM Card or Embedded eSIM" of chapter "Ingredients for NB-IoT Design Concepts").

In fact, IoT data traffic is different from data transferred from/to mobile phones or smartphones. By nature of IoT, only small amount of data is being transferred, infrequently and mainly in uplink direction. For MVNOs this is a good opportunity to differentiate and offer tailored pricing for typical IoT usage scenarios. MVNO EMnify, for example, bills IoT data traffic based on a granularity of 1 kB (although TCP protocol overhead is also included, i.e., added on top of payload data).

MVNO subscription plans are also offered by some manufacturers of cellular network modules. This might be an interesting option for application developers who have shortlisted Telit or Sierra Wireless as potential suppliers (see section "Vendor Overview" of chapter "Ingredients for NB-IoT Design Concepts"). u-blox offers a dedicated service for MQTT messaging. Many MVNO offer SIMs and/or eSIMs including cloud tools for monitoring device traffic and for SIM management, e.g., to activate or deactivate a SIM or change eSIM profiles. Another common option is to incorporate device data from MVNO tool into customer IoT applications via a RESTful API, i.e., via HTTP commands.

MVNOs specialized on IoT business are listed here (in alphabetical order):

- 1NCE: https://1nce.com
- EMnify: www.emnify.com
- Monogoto: https://monogoto.io
- Netmore: https://www.netmorem2m.com
- Onomondo: https://www.onomondo.com
- Ritesim: https://ritesim.com
- Soracom: https://www.soracom.io
- Twilio: https://www.twilio.com
- UROS Connect: https://connect.uros.com

Designing an NB-IoT Device

As a starting point for project-specific designs and evaluations a few sample system implementations are provided in this book. In this section we introduce two design concepts for NB-IoT devices, each is representing a typical NB-IoT target application group (see section "NB-IoT Use Cases" of chapter "IoT Target Applications"):

- **Monitoring/Detection**: Environmental Sensor with 10+ years battery lifetime. This device is designed for stationary operation. It is supposed to transmit local IoT data (temperature, humidity, pressure) periodically as scheduled. Push data model.
- **Track/Localize**: Object Localizer. This device is a battery-powered mobile unit attached to a host object for theft/loss protection purposes. It will provide current geo position on request, i.e., when getting paged. Pull data model.

These sample designs are representing both major application areas identified for NB-IoT cellular technology and both operation modes, one for periodic **push**, the other for **pull** of IoT data.

RasPi Mockup and Network Tester

Electronic design professionals can leverage worldwide "maker" trend and use popular Raspberry Pi computer board as a test tool or as an IoT device mockup. Raspberry Pi (short: "RasPi") is a small versatile play-and-play system which works out-of-the-box and offers common interfaces for users and peripherals. For operation, just a standard USB keyboard and mouse plus HDMI display are needed. RasPi comes with an onboard Debian GNU/Linux OS and many generic software ingredients for all kind of embedded computing applications. Customers can add application-specific circuitry via GPIO pins or use I²C bus for IoT peripherals like sensors, etc.

© The Author(s), under exclusive license to Springer Nature Switzerland AG 2022
K. Heins, *NB-IoT Use Cases and Devices*, https://doi.org/10.1007/978-3-030-84973-3_5

For NB-IoT trails, company sixfab is offering a RasPi add-on board called "Cellular IoT HAT" (HAT stands for "Hardware Attached on Top"). It is based on Quectel's BG96 network module (URL: https://sixfab.com/product/raspberry-pi-lte-m-nb-iot-egprs-cellular-hat) and offers GSM/GPRS as a secondary connectivity option. For user convenience and immediate start, a versatile SIM card by "virtual" network operator Twilio is included (see section "Roaming and MVNOs" of chapter "NB-IoT Network Deployment"). Here, we just use it as an NB-IoT network tester and demonstrate use of AT command interface for network operations. Figure 1 shows the core setup with RasPi, sixfab board, and antenna.

The sixfab NB-IoT RasPi extension is easy to install and comes with a low-level library for hardware control and some sample Python programs. Besides a basic understanding of RasPi user interface, Quectel's BG96 AT Commands Manual [22] will be the starting point for own network tests and evaluations.

The following test program contains a sequence of AT commands which allows to determine availability of a suitable NB-IoT connection. As a practical example, this short Python application has been used for a real-world connection attempt at a specific location near Munich in Germany. Two libraries are imported:

Fig. 1 RasPi NB-IoT Network Tester

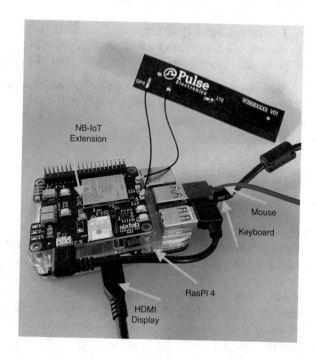

```
#! /usr/bin/env python3
from cellulariot import cellulariot
from time import sleep
node = cellulariot.CellularIoT()
```

Key software component is library **cellulariot.py** provided by sixfab. It is using the standard Python module **pyserial** which is providing access to the RasPi serial port. It is used to implement an AT command interface between a RasPi application and the UART interface of the BG96 modem function. As a reference, source code of **cellulariot.py** is provided here: https://github.com/sixfab/Sixfab_RPi_CellularIoT_App_Shield/tree/master/cellulariot.

Library function **setupGPIO()** is used to control some RasPi GPIO pins which are used to power up and enable the BG96 module. See Cellular IoT HAT board schematics for details, URL is https://github.com/sixfab/Sixfab_RPi_CellularIoT_App_Shield/blob/master/schematics/Sixfab_RPi_CellularIoT_App_Hat_Schematic.PDF.

```
node.setupGPIO()
node.disable() #GPIO pin BG96_ENABLE
sleep(1)
node.enable() #GPIO pin BG96_ENABLE
sleep(1)
node.powerUp() #GPIO pin BG96_POWERKEY
```

Key library function is **sendATComm()** which can be used to send an AT command to the BG96 module. The original AT command (from [22]) must be converted according to syntax specified in **cellulariot.py**.

AT command **QCFG** determines some details of the network selection procedure (see [22] for detailed description and parameters). In order to speed up the network search process, BG96 module can be configured to focus on specified options, irrelevant options can be excluded. Parameter **nwscanmode** can be used to select allowed RATs (RAT = Radio Access Technology), parameter **iotopmode** specifies relevant network categories. BG96 supports LTE Cat. M, LTE Cat. NB1 and GSM/GPRS. For this example, search is limited to NB-IoT networks only:

```
node.sendATComm("AT+QCFG=\"nwscanmode\",3,1","OK\r\n")
node.sendATComm("AT+QCFG=\"iotopmode\",1,1","OK\r\n")
```

Parameter **nwscanseq** can be used to specify the search order. For example, it might be useful to look for NB-IoT first, but—if not available—continue to search for a GSM/GPRS network as a fallback (parameter would be "**0301**"). Here, search is limited to NB-IoT ("**03**"):

```
node.sendATComm("AT+QCFG=\"nwscanseq\",03,1","OK\r\n")
```

In this particular example, certain LTE frequency bands were irrelevant because only B3, B8, and B20 have been assigned for NB-IoT deployments in Europe:

```
node.sendATComm("AT+QCFG=\"band\",F,80084,80084,1","OK\r\n")
```

After configuration of search and selection criteria, the BG96 module will independently scan available cells accordingly. By nature, for NB-IoT this will take some extra time because of its narrow bandwidth (200 kHz) and allowed weak network coverage of up to 164 dB MCL incl. assignment of CE levels 0–2 for an allocated channel. See [23] for Quectel BG96 test results and recommendations. Applied 60 s waiting time might not be sufficient in all cases.

```
sleep(60) # Allow 60 seconds for network scan
```

At our test location, connection to a suitable NB-IoT network was successful. AT command **QNWINFO** ("Query Network Information") returned.

```
node.sendATComm("AT+QNWINFO","OK\r\n")
+QNWINFO: "CAT-NB1","26203","LTE BAND 20",6152
```

indicating PLMN = 26203 which is referring to network operator Telefonica (i.e., O2 in Germany).

Note: Using a universal SIM provided by MVNO was very useful in this particular case because it works with many MNOs offering NB-IoT connectivity. For example, a Vodafone SIM would not work for selected test location.

Additional AT command **QCSQ** ("Query and Report Signal Strength") returns

```
node.sendATComm("AT+QCSQ","OK\r\n")
+QCSQ: "CAT-NB1",-76,-77,205,-3
```

i.e., RSSI = −76 (RSSI = Received Signal Strength Indicator)
RSRP = −77 (RSRP = Reference Signal Received Power)
SINR = 205 → 205/5 = 41 dB (SINR = Signal to Interference plus Noise Ratio)
RSRQ = −3 dB (RSRQ = Reference Signal Received Quality)

In addition, AT command **AT+QCFG="celevel"** ("Get LTE Cat NB1 Coverage Enhancement Level") returned "0" indicating that for selected test location no repetitions would be applied to data transmission between NB-IoT device and network base station.

```
AT+QCFG="celevel"
+QCFG: "celevel",0
```

In general, our RasPi/NB-IoT system can be used by for various studies and for deployment planning in combination with coverage plans provided by potential or selected MNO partners. In addition, it might be used as a mockup for project preparation or demonstration purposes. And, last but not least, it provides an opportunity to collect hands-on experience with NB-IoT cellular networks for preparation of own projects.

Note: This short application program is called **nbiotscan.py** can be downloaded from URL www.iot-chips.com/springer. Password is "narrowband."

Low-Power Device Design

In general, power consumption of an NB-IoT device is closely related to NB-IoT network performance and does not depend much on selection of suitable electronic components (modem, MCU, sensors, etc.) for device design. Key aspects are network configuration and individual device location which are impacting transmission time and latency. Several scientific studies are dealing with these subjects. See section "Latency" of chapter "Cellular IoT Technology" and sources provided in "References."

Main driver for power consumption is the RF part of the modem of an NB-IoT device, when in use during data transmission or reception periods. Uplink data transmission current consumption is around 250 mA, cost of downlink transmission is just 50 mA. The IoT application program can schedule transmission of data packets in combination with NB-IoT power saving mechanisms like eDRX and PSM according to use case requirements. This way, some optimizations can be done in an effort to reduce power consumption. For example, the embedded MCU may perform calculations with 2 mA and perform interactions with peripheral components (e.g., sensor) in parallel to a modem in PSM state while consuming only few microamps during this period. The MCU will wake up the modem infrequently only and transmit consolidated data only once per day, instead of once per hour. This kind of scenarios can be calculated and compared with our "NB-IoT device battery lifetime calculator" which will be introduced later.

Besides these user controllable aspects, an optimized use of the NB-IoT network is a key for battery-powered NB-IoT devices. Unfortunately, network behavior depends on network owner (MNO) and configuration of relevant parameters (e.g., configuration of CE thresholds) might even change from one cell to another. Some timer settings (e.g., TAU timer T3412 or Active Timer T3324, see illustration in Fig. 13 of chapter "Cellular IoT Technology") are configurable, others are not known by users. For example, a user does not know how many data transmissions (repetitions) will be applied by the network to a device located in a lossy network coverage condition. For NB-IoT projects relying on a certain minimum battery lifetime for all devices in the field, this unpredictability is a challenge because device power consumption may not be the same for all deployed devices, so it cannot be calculated from scratch. Instead, power consumption of each NB-IoT device will depend on its individual location and attached network cell. Own studies and local verification (see section "RasPi Mockup and Network Tester") and individual optimization during local installation of an IoT device at its final destination would help (see section "Local Device Setup"). But independent from local network configuration parameters, there are a number of design aspects which are worthwhile to consider right from the beginning.

Manage Immediate RRC Release

Most important for long battery life is to switch off radio if not needed. As long a user device is in RRC_Connected mode it will be receiving all base station signaling. While receiving, the average power consumption of a modem is around 45 mA. Unfortunately, the user device cannot end RRC_Connected state by itself. Instead, it has to wait the network inactivity timer to expire even if no further transmission is planned of expected by the device (see paragraph Fig. 13 of chapter "Cellular IoT Technology"). If the RRC connection is kept alive each time for another few seconds without need, this will cost a significant amount of battery capacity over time. For example, 45 mA power for a duration of 5 s each day will result in a total battery capacity of 22.8 mAh per year (This calculation has been done with our "NB-IoT device battery lifetime calculator").

To avoid this waste of energy, the RAI feature can be used, see section "Release Assistance Indication (RAI)" of chapter "Cellular IoT Technology." RAI has been released in 3GPP Release 13 (for non-AS) resp. Release 14 (for AS). For example, for u-blox SARA-N3 cellular module the following AT command can be used to enable RAI [17]:

```
AT+MIPLSETRAI=<rai_mode>
```

with

```
rai_mode=1:
```

release the connection after the uplink data is sent

```
rai_mode=2:
```

release the connection after the first data is received on downlink

Note: If RAI is not supported by local network(s), user devices can use a workaround by sending command **AT+CFUN=0** to the modem which disables both transmit and receive RF circuits during RRC_Connected state until the network has initiated RRC_Idle state.

Optimize Downlink Traffic Scheduling

Each time an RRC connection has been released, the device starts to perform paging cycles using DRX or eDRX power saving mechanisms. The Active Timer T3324 controls the length of this period during which the device is reachable by the network that is started immediately after the NB-IoT device releases the RRC connection and enters the idle state. The value of T3324 can be in the range of 0 s to roughly 3.1 h. After this timer expires, the device enters PSM, i.e., ultimate low-power consumption.

T3324 timer value greatly affects battery lifetime. Apparently, the lower the timer value, the faster the modem will go into PSM state. In most networks, can be set to ZERO to immediately enter PSM. As a consequence, the device would not be available for downlink traffic during RRC_Idle states at all. This configuration option might be appropriate for some uplink-focused IoT applications, e.g., for Remote Monitoring/Detection which are designed for scheduled (i.e., planned) download traffic only. In this case, active listening to paging occasions would be executed in RRC_Connected mode only.

In 3GPP Release 15 a wake-up signal (WUS) feature has been introduced to support optimization of download traffic, see section "Wake-Up Signal (WUS)" of chapter "Cellular IoT Technology." This signal is used to notify a specific device about relevant paging messages to come. If WUS is present, the device should monitor NPDCCH for paging, otherwise it allows the device to skip next paging occasion and keep receiver in idle mode. Obviously, WUS feature requires support by network as well as by modem hardware, so it will not be usable by every NB-IoT project.

In general, designers of NB-IoT applications should carefully manage downlink traffic in a way that a battery-powered device has to listen only to infrequent and short paging windows, if possible.

Avoid Coverage Enhancement Levels CE1 and CE2

NB-IoT supports devices at locations with bad signal strength with up to 164 dB coupling loss. This feature works with repeated data transmissions (up to 128 UL resp. 2048 DL repetitions) for affected user devices (see section "Coverage Extension" of chapter "Cellular IoT Technology"). This is beneficial for deep indoor penetration or coverage of wide rural areas, but in return these repetitions have a negative impact on data rate, latency, and power consumption. Related studies [16] have shown that coverage enhancement levels for devices facing coupling losses of >144 dB are causing significantly extended data transmission time with tremendous impact on device power consumption resp. battery lifetime. Consequently, for battery-powered NB-IoT devices, CE levels >0 should be avoided.

A CE level (incl. number of applied repetitions) is assigned by a cell to each device individually. This information is part of the Downlink Control Information (DCI) provided by the eNodeB. DCI is 23 bits long and contains downlink or uplink scheduling information, or uplink power control commands for a specific RNTI (temporary device ID). DCI message format N0 is associated with the NPUSCH uplink channel which is used for device payload data. Once the user device has obtained a DCI N0 message, bits 17–19 are indicating the repetition number I_{REP}:

1	2	3	4	5	6	7	8	9	10	11	12	13	14	15	16	17	18	19	20	21	22	23
0																I_{REP}						

Repetition number I_{REP} is converted into number of applied repetitions N_{REP} as follows:

I_{REP}	0	1	2	3	4	5	6	7
N_{REP}	1	2	4	8	16	32	64	128

For NB-IoT application developers who do not have access to low-level DCI messages and who are using off-the-shelf network modules (see section "NB-IoT Cellular Network Modules" of chapter "Ingredients for NB-IoT Design Concepts") dedicated modem commands might be available to determine assigned CE level. For example, for u-blox SARA-N3 module [17] the following command

```
AT+CRCES
```

will return the coverage enhancement status as

```
+CRCES: <AcT>,<CE_level>,<CC>
OK
```

with **<CE_level>= 0...3** indicating the assigned CE level.

Depending on returned coverage enhancement status the NB-IoT device might decide to refrain from transmitting data and, for example, look for a different cell in reach.

These AT commands are proprietary. For Telit ME910G "Reading Coverage Enhancement Status" command is the same [18]. But Quectel BG96 module [22] offers a similar function with a different command syntax: **AT+QCFG="celevel"**

Optimize Uplink Transmission

In general, transmission power is one of the most critical parameters for battery lifetime. To minimize this impact, for Release 13 NB-IoT user devices two **power class** options could be used: power classes 3 with 23 dBm and power class 5 with 20 dBm. 3GPP Rel 14 adds another power class with 14 dBm:

Power class	Output power (dBm)	Output power (mW)
3	23	200
5	20	100
6	14	25

Besides lower power consumption for uplink transmission, a lower power class will allow manufacturers to integrate of the power amplifier into a single-chip implementation at lower cost and for IoT devices with smallest form factor. On the other hand, lower power classes are challenging network operators as well for NB-IoT users. By nature, reduced output power of an IoT device will result in less transmission range, i.e., a network supporting lower power classes will suffer from a reduction of coverage. Extended-range NB-IoT cells working over 100 km distance might not be achievable with power classes 5 or 6. For user devices in the field this means that it will enter Coverage Enhancement levels CE1 or CE2 sooner than a power class 3 device. Associated repetitions of uplink data transmissions will cost extra power and might exceed energy saving gained through reduced output power. As a consequence, potential benefit of a lower power class significantly depends on individual device location, thus it should be considered case-by-case separately for each device individually.

In general, transmission power is adjusted internally by the modem according to allocated network resources (bandwidth) and actual path loss. Please refer to 3GPP TS 36.213 [11] for further information about uplink power control. But some modems offer proprietary user functions allowing more visibility and control. For example, for u-blox SARA-N3 module [17] AT command `AT+NPOWERCLASS=<power_ class>` can be used to change power class of an NB-IoT device. For mentioned SARA-N3 module, new power class 6 (14 dBm) is listed as an allowed input parameter. But power class 6 might not be implemented by all NB-IoT modems (yet).

For user devices compliant with 3GPP Rel 15 working in an up-to-date NB-IoT network, EDT (see section **"Early Data Transmission (EDT)"** of chapter "Cellular IoT Technology") is an excellent power saving mechanism because transmission of small data packages (<1000 bits) can be handled during an RRC connection procedure. In this case, the device does not have to transit to RRC_Connected state for uplink transmission.

Matching of Antenna

Antenna matching is particularly important for NB-IoT designs, esp. for battery-powered devices. Transmission power is one of the most critical parameters for low-power consumption of an NB-IoT device, it should be as low as possible resp. power should be used as efficiently as possible. In fact, transmitters are typically designed to feed power into a pure resistive load of 50 Ω, i.e., without any capacitive or inductive reactance. An impedance mismatch will reduce transmission power radiated by the antenna, the reflected current can overheat transformer cores and cause signal distortion. By nature, the antenna and feedline impedance vary with frequency of the transmitted signal.

Maximum power is transferred only if matching is done at the antenna in conjunction with a matched transmitter and feedline, producing a match at both ends of the line. So, in order to optimize power transfer between a radio transmitter and its antenna a suitable circuitry that corrects the mismatch has to be added in between. The most straightforward matching-network topology is called the L network. This refers to eight different L-shaped circuits composed of two capacitors, two inductors, or one capacitor and one inductor. Figure 2 illustrates one out of eight possible configurations.

Depending on location of deployment (country, network operator), NB-IoT devices will have to be prepared to use several different frequency bands, see "NB-IoT Network Deployment." This means that a suitable **matching network for each NB-IoT carrier frequency** has to be available by design for all possible options— and to be activated accordingly at runtime. This can be done by use of off-the-shelf multi-port RF switches which are controlled via GPIOs of the NB-IoT network module. See our "Design Concept #2: Object Localizer" which is addressing this subject for this particular project.

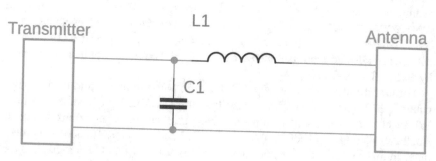

Fig. 2 Antenna Matching Network (sample configuration)

Utilize Sleep Modes

Quite obviously, selection of low-power components for an NB-IoT device design is key esp. for battery-powered devices. This kind of NB-IoT device will be sleeping most of the time, same applies to all other components as long as they do not have to sense a local parameter permanently. For an NB-IoT design, PSM modem state can trigger all other active devices to follow into low-power mode, if possible. Besides the PSM mode of the modem itself, low-power consumption of host MCU and sensors are major contributors to a long battery lifetime.

For example, looking at currently available NB-IoT Cellular Network Modules, best candidate offers lowest value of PSM input current of just 0.5 µA (u-blox SARA-R5), worst is at 22 µA. In fact, both values appear small at first glance, but might significantly reduce product lifetime. For example, battery lifetime of our "Design Concept #2: Object Localizer" has been calculated with 14.4 years based on PSM input current of 2.7 µA (Nordic Semiconductor nRF9160). If we use Sierra Wireless WP7700 module with 22 µA instead, we end up with an estimated lifetime of 7.9 years, i.e., PSM input current is cutting total product lifetime almost by half in this case.

Estimate Overall Power Consumption

Successful design of a low-power NB-IoT application incl. associated devices requires a close look at all components and how they cooperate. And, as a general principle, it will be important to keep power consuming activities as short as possible. There are many good reasons to reduce energy consumption, lower operational cost is one of them. Especially for application scenarios with battery-powered devices, low device cost and low operational cost are key selling points.

In fact, for **zero-touch IoT devices** using a pre-installed battery with fixed capacity and no option to recharge or to replace it, **battery lifetime is identical with product lifetime**. This means that the product has reached its end of lifetime and will be taken out as soon as the battery does not power up the device any more. Consequently, battery runtime has to exceed targeted product lifetime, e.g., 10 years. In an effort to estimate required minimum battery capacity, we need to understand power consumption characteristics of all involved components and relevant aspects of the NB-IoT network. Then, we can list all capacity-eating elements (duration, power consumption) and consolidate them.

This will be straight-forward for electronic components used for IoT device hardware because specifications (e.g., section "Electrical Characteristics" in product data sheet) will provide required technical information. But some NB-IoT network configurations have significant impact on device power consumption, too. Unfortunately, many of them are unknown and out of user control, e.g., some timer settings or applied output power. In particular, it is not possible (for a network user) to quantify uplink data transmission periods incl. overhead like synchronization, RRC connection/release, random access, etc. Many network configurations are MNO-specific and might even differ from cell to cell. This is why you cannot calculate power consumption precisely from scratch. Consequently, an IoT project should carefully select a suitable network partner und consider to perform own tests in target deployment area(s). See also section "RasPi Mockup and Network Tester."

Good news for IoT application developers is that modem manufacturers are providing helpful tools and higher-level AT commands for better visibility and network control. For example, for u-blox SARA-N3 module [17] function **AT+NEWSTATS[=<type>]** can be used to retrieve most recent operational statistics of the module. Depending on the <type> parameter the information text response provides different information as radio specific, cell, application core memory, block error rate or throughput information. For power consumption analysis, **AT+NEWSTATS="RADIO"** will return duration of last downlink and uplink transmissions and applied output power, for example.

In addition, worst case assumptions and best guesses can be used for initial calculation of radio time, e.g., originated from studies like [24]. In this paper, behavior of a real user device has been evaluated in cooperation with a real-world network implementation (Vodafone NB-IoT in Barcelona/Spain, Band 20). In particular, energy consumption has been measured based on different device configurations and network dependencies. For example, current traces are illustrating the effect of the network inactivity timer keeping the device in RRC_Connected state for another 20 s (!) after last transmission, performing DRX cycles during this period before the device is released into PSM.

Another measurement ([24] Fig. 3c) is showing the effect of a dedicated setting for IoT use cases like water meters or environmental sensors which are focusing on uplink data reports with a limited need for download traffic and associated paging messages. In this case, the connection is released immediately after data upload, see section "Release Assistance Indication (RAI)" of chapter "Cellular IoT Technology." In addition, T3324 Active Timer has been set to zero and disabling subsequent DRX cycles in RRC_Idle mode so that the device can rapidly enter PSM mode. This setting results in lowest radio power consumption.

In order to estimate expectable power consumption of IoT deployed devices during field operation and required battery capacity to be installed, the following calculation spreadsheet can be used. For this purpose, we need to enter all power-relevant data. Unfortunately, we are facing **significant NB-IoT network variability**, and there are only few instruments available for users to control applied network configuration and related power consumption. This is leading to poor predictability of a NB-IoT network behavior, esp. for uplink data transmission timing. At least for this part, a decent calculation of expectable power consumption is not possible, so we have to use best guesses as a starting point. Further evidence and more precise input data can be delivered by dedicated real-world measurements in the field which will finally improve calculation result and related design decisions.

NB-IoT Device Battery Lifetime Calculator

Our spreadsheet is calculating the average power consumption of an NB-IoT device—based on relevant activity data incl. duration, event frequency, and power consumption. These inputs will be consolidated, and in combination with a pre-selected battery capacity which has been installed at deployment time, the calculation will return the overall battery lifetime. The calculation scheme assumes that used NB-IoT device will be sleeping most of the time until wake up for an activity whenever required. This means that all relevant activities must be considered first, then—for the remaining time—energy consumption of components in their default power-down mode will be added automatically by the calculator.

The idea is to observe the impact of changed communication scenarios (e.g., more uplink data, but less frequently) or impact of a different component selection with other characteristics or power saving modes. We will feed this tool with application-specific input data for our two design concepts in the following sections "Design Concept #1: Environmental Sensor" and "Design Concept #2: Object Localizer."

For following tutorial, we have entered five sample main activities: installation, modem TX radio, modem RX radio, sensor measurement, and host MCU activity. For each we need applicable power consumption and duration of each single event resp. number of occurrences. All fields filled with orange background (for printed version: light gray) or green need to be filled (Fig. 3):

- **Current (mA)**: from datasheet for particular operation mode and selected parameters, e.g., for a data transmission event: modem output power and selected frequency band.
- **Duration of a single event (ms)**: e.g., enter "2000" for 2 s to be considered for a single occasion.
- **Additional periodic occasions per day**: this field is used for multiple executions per day. Enter "23" if activity should be considered 24 times per day (i.e., if executed every hour), "0" means that no additional occurrences are planned, so this event will be considered once in the total result.
- **Periodic occasions per year**: enter "365" if activity should be considered 365 times per year (for daily execution), enter "730" for execution twice a day. If "0" is entered, no periodic execution is calculated (only single occasions, if entered).
- **Single occasion(s)**: used if event occurs only once during product lifetime, e.g., an initial device configuration after deployment.

Then, after all relevant current-consuming activities have been entered, we will have to add average consumption (in mA) for each device component when running in its low-power mode, e.g., for modem resp. cellular network module, local MCU, sensor, etc. The calculation will exclude active periods which have been entered in same spreadsheet before, i.e., power-down consumption will be applied for remaining time only.

Calculation results will be displayed in fields with blue background with white characters (printed version: dark gray background) labeled as "periodic capacity

Item	current (mA)	duration of single event (ms)	additional periodic occasions per day	total duration per scheduled day (ms)	total duration per day (hrs)	periodic occasions per year	total duration per year (hrs)	consumed capacity per hour (mAh)	periodic capacity consumption per year (mAh)	single occasion(s)	add capacity consumtion for single occasions (mAh)
installation procedure (flat, one-time)	250	400000	0	400000	0,111111	0	0,000	0,00000	0,00000	1	0,1111
TX radio @20dB	200	2000	0	2000	0,000556	365	0,203	0,00463	40,55556	0	0
RX radio	45	2000	1	4000	0,001111	182	0,202	0,00104	9,10000	0	0
sensor measurement	45	2000	0	2000	0,000556	365	0,203	0,00104	9,12500	0	0
MCU activity	20	5000	0	5000	0,001389	365	0,507	0,00116	10,13889	0	0
total ACTIVE time	560			413000	0,114722		1,115	0,007868	68,92		0,1111
modem in power save mode (PSM)	0,005										
sensor idle state	0,005										
MCU idle	0,01										
total IDLE time (= 100% minus active time)	0,02			85987000	23,885278		8758,9	0,02000	175,18		
Subtotal periodic consumption									244,10		
TOTAL				86400000	24,000000		8760,00	0,02786			244,21
				86400s * 1000 = one day			8760hrs = one year				⇩
battery capacity (mAh)	3000								battery lifetime in years:		12,3

Fig. 3 Sample battery lifetime calculation

consumption per year in mAh" resp. "add. capacity consumption for single occasion(s) in mAh." Based on installed capacity value to be entered in orange field (printed version: light gray background), expected "battery lifetime in years" will be presented in bottom right corner field.

Note: This calculator has been implemented as an Excel spreadsheet and can be downloaded from this URL: www.iot-chips.com/springer. Password is "narrowband."

Design Concept #1: Environmental Sensor

Remote monitoring is a typical IoT application (see section "NB-IoT Use Cases" of chapter "IoT Target Applications"). As an example, an environmental authority or site management wants to track local ambient air conditions of a remote location where no power outlet is available and no local network can be used. For this use case, an IoT device should be powered by battery and a cellular network should be used in order to ensure reliable operation. In addition, the IoT device should be prepared for zero-touch operation during complete device lifecycle, i.e., it should be 100% **maintenance-free** and should not need any battery replacement or recharge until its end of lifetime. This means that the IoT device should be designed for maximum efficiency in order to reach 10 years battery life. NB-IoT particularly has been specified to meet this kind of requirements.

Principle of Operation

In order to achieve maximum battery lifetime, proper orchestration of power management features of all three main components is required: network module, host MCU, and sensor. Master role is alternating between two of them: IoT application program is executed by the host MCU, but wake-up management is handled by the cellular module in cooperation with the NB-IoT network via V_INT pin indicating modem in PSM mode, see block diagram Fig. 4.

For this design concept a Bosch BME280 **environmental sensor** is used. But same concept also works fine for other "Remote Monitoring/Detection" use cases with other low-power devices as long as standard interfaces like I²C or SPI or GPIOs are used. For example, you can use a smoke detector or fluid level meter or

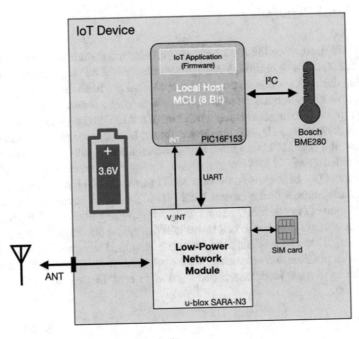

Fig. 4 Environmental sensor design—block diagram

time-of-flight sensor for tracking presence of an object (see section "Sensors" of chapter "Ingredients for NB-IoT Design Concepts" for further information and selection criteria).

Our IoT device is based on a u-blox SARA-N3 **NB-IoT cellular module** which is specialized on Cat. NB2 networks (3GPP Rel. 14) and offered for lowest price. For communication with a network cell, a **matched external LTE antenna** must be connected (see section "Matching of Antenna"). The device is powered by one or two 3.6 V Li-ion 18650-sized batteries allowing flexible configuration and **total capacity of a few 1000 mAh up to several Ah.**

For this IoT device design a discrete **host MCU** is used, a simple 8-bit MCU, for example, a low-power and low-cost PIC16F15313 with 3.5 kB onboard flash memory in a small 8-pin SOIC package. This MCU controls operation of the cellular network module via AT command API (see section "AT Command Interface" of chapter "Ingredients for NB-IoT Design Concepts") and executes customer-defined IoT application software, e.g., to manage function of the peripheral sensor. In our case, the sensor (temperature, humidity, pressure) is connected to the I²C bus of the host MCU.

PIC16F15313 has an internal clock source of 32 MHz with a user-configurable system clock divider allowing to balance power consumption vs. performance according to application requirement according to specified 32 µA/MHz@1.8 V during operation.

Functional Description

Figure 5 is illustrating the device operation from a software point of view. After individual local configuration, each deployed device will be prepared for zero-touch operation during its complete product lifecycle and run in an infinite program loop.

Note: Numbers in brackets refer to program steps. Example: "(0)"

After delivery to target location, power up of the IoT device or use of reset button resets all components (1), and the device should be ready for **manual installation** (2). Then, the host MCU should be able to communicate with the network module (aka "modem") via UART interface and configure it according to application requirements (3). In particular, the host MCU firmware will have to make sure that the module connects to the closest local NB-IoT network tower covered by subscription plan of chosen MNO partner and corresponding SIM card. This process is important because distance to cell tower will determine device power consumption (see section "Estimation of Battery Lifetime"). Required transmission power should be as low as possible and network module configuration should be adjusted accordingly during device installation. See section "Local Device Setup" for further information.

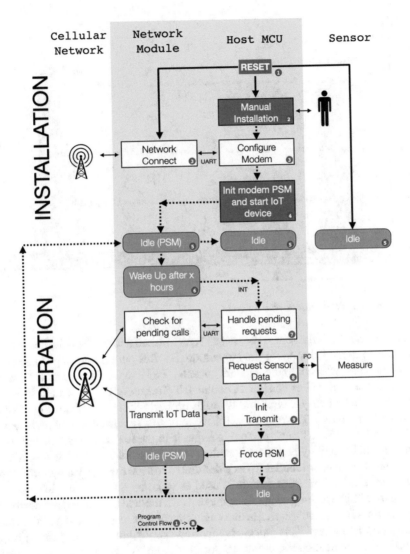

Fig. 5 Environmental sensor—application program flow

Our design concept is based on the idea that all local IoT actions are handled during a single periodic activity time slot (see Fig. 6). After installation (i.e., after steps (3) and (4)), the IoT device program will loop endlessly, no deviation from this flow is planned, and IoT device will not be reachable during PSW periods.

Note: If needed, the embedded IoT application (on host MCU) can be prepared to interact at any time, i.e., to react to incoming network messages and wake up the device immediately. For this purpose, a customer design can take advantage of dedicated pin CTS/RI (RI = "ring indicator," see Fig. 8) to submit an interrupt signal to the host MCU.

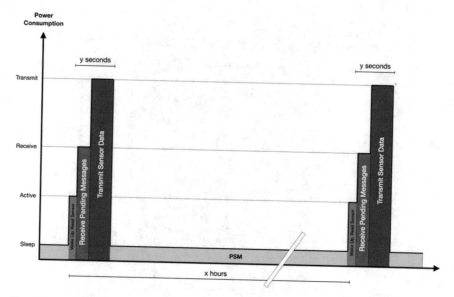

Fig. 6 Environmental sensor—activity cycle

As soon as the initial setup process has been finished and verified, the host MCU can prepare the IoT device for field operation. For this purpose, in step (4) a set of AT commands will configure modem to enter PSM mode (see section "Power Saving Mode (PSM)" of chapter "Cellular IoT Technology"), to remain in PSM mode as a default (5 μA typ.) and to wake up periodically automatically after **x hours** (configurable). The module will have to negotiate related timing parameters with connected NB-IoT network so that requests to the IoT device will be buffered during agreed inactivity periods. When done, cellular module as well as host MCU will **enter power-down modes (5) as a default condition during product lifetime**.

In fact, the NB-IoT PSM feature is used to wake up the complete IoT device. Configured PSM wake-up schedule of the network module (6) will trigger **periodic activity slots** for transmission of actual environment data to a pre-configured recipient, e.g., an IP address. This process is repeated every **y hours** (configurable) resp. according to application requirements. And it should be kept as short as possible because power consumption will dramatically increase when radio is turned on. Second intention is to use this periodic activity period as an opportunity to check for incoming messages. The idea is to use them to update device configuration data, e.g., network parameters (see Fig. 6), if required.

SARA-N3 module is featuring a PSM indicator pin (V_INT) which can be used to wake up the local **host MCU** via interrupt INT pin (6) which is then **taking control** of all further actions during IoT device active period. As a first action, the local MCU will handle pending network requests to the IoT device, if any (7). Then, local host will **initiate a single measurement** via I²C bus (8). This request will wake up BME280 from default sleep mode (0.1 μA) and takes around 1 s to execute. As soon as sensor measurement data has been received, host MCU will submit AT commands to request transmission of IoT data according to selected protocol, e.g., MQTT (9).

Now, for this activity slot, all planned tasks have been executed, and the IoT device is ready to re-enter idle mode until next PSM wake-up event will occur. For this purpose, host MCU will reactivate module PSM mode (A) and return to sleep itself (B). Then, application program loop will restart at (5) and wake-up management will return to the network module.

Estimation of Battery Lifetime

In general, keeping activity periods as short as possible is a key for lowest power consumption. For our component selection, even during active periods, power consumption is moderate:

- PIC16F15313: 8 µA @ 32 kHz (internal clock) resp. 32 µA/MHz after reduction of clock frequency
- BME280: 3.6 µA

In order to estimate device power consumption and battery lifetime we use **our "NB-IoT device battery lifetime calculator"** and enter design-specific data. Most power is required by SARA-N3 in active mode (6 mA) and esp. when radio is turned on. During receive mode, current consumption will be 46 mA. If network signal strength is good enough, attached network can advise the NB-IoT device to reduce TX output power, but worst case assumption for our power consumption calculation is power class 3 (23 dBm) which is 220 mA for SARA-N3 module. Manual installation (see section "Local Device Setup") will have to make sure to avoid CE level 2, i.e., number of repetitions of uplink data transmission and corresponding delay should be small, see section "Latency" of chapter "Cellular IoT Technology." Normally, **attached network cell will not change** during product lifetime, this means that standard NB-IoT "RRC Suspend/Resume" feature can be used after modem wake-up from PSM, i.e., no negotiation of connection parameters will be required. After BME280 wake-up, it will take 1 s to perform measurement, read sensor data, and transmit to network, so we should end up with a duration of no more than $y = 5$ s for each this task, see Fig. 6. The usage scenario for our calculation is: sensor data is being transmitted **twice a day**, i.e., every 12 h.

In addition, we have to consider occasional reception of pending messages containing device updates. This might happen 10 times during product lifetime and might take 10 s incl. MCU processing. These two activities and are entered in our calculation spreadsheet, see Fig. 7.

1. 5 s TX activity with 220 mA, two times periodically each day
2. 10 s RX activity with 46 mA, ten single occasions

Item	current (mA)	duration of single event (ms)	additional periodic occasions per day	total duration per scheduled day (ms)	total duration per day (hrs)	periodic occasions per year	total duration per year (hrs)	consumed capacity per hour (mAh)	periodic capacity consumption per year (mAh)	single occasion(s)	add capacity consumtion for single occasions (mAh)
periodic wake-up and TX activity @23dBm	220	5000	1	10000	0,002778	365	1,014	0,02546	223,05556	0	0
occasional device update (RX activity + MCU)	46	10000	0	10000	0,002778	0	0,000	0,00000	0,00000	10	0,0278
	0	0	0	0	0,000000	365	0,000	0,00000	0,00000	0	0
	0	0	0	0	0,000000	365	0,000	0,00000	0,00000	0	0
total ACTIVE time	266			20000	0,005556		1,014	0,025463	223,06		0,0278
SARA-N2 module PSM mode	0,005										
BME280 idle mode	0,001										
PIC MCU idle	0,001										
total IDLE time (= 100% minus active time)	0,007			86380000	23,994444		8759,0	0,00700	61,31		
Subtotal periodic consumption									284,37		
TOTAL				86400000	24,000000		8760,00	0,03246			284,40
				86400s * 1000 = one day			8760hrs = one year				⇩
battery capacity (mAh)	3000								battery lifetime in years:		10,5

Fig. 7 Environmental sensor—battery lifetime

In addition, we have to enter low-power mode consumptions of all active components:

- SARA-N3: 5 μA typ. (PSM mode)
- PIC16F15313: below 1 μA
- BME280: below 1 μA

For our estimation we consider use of a **Li-ion battery with 3000 mAh**, so for selected use case schedules we end up with an estimated battery lifetime of **10.5 years**.

Of course, schedules can be adjusted according to application requirements. Depending on activity frequency, power consumption will change. But, in any case, our design offers an assembly option for one extra 18650 battery which allows to increase total battery capacity to several Ah.

Local Device Setup

Device deployment is kind of challenging because for this particular use case you probably want to avoid network cell assignment of CE level 1 or 2 to the device (see section "Low Power Device Design" of chapter "Designing an NB-IoT Device"). But exact cell positions and network coverage at each target location might not be known or not 100% reliable. CellMapper tool or a dedicated NB-IoT network tester might help to optimize individual device location (refer to chapter "NB-IoT Network Deployment"). But in any case, an individual manual configuration and activation of each IoT device at its target location might be required. This manual installation procedure should

- Identify local network availability and options,
- Attach device to the most appropriate cell,
- Finetune device location, and
- Start field operation of the device.

For this purpose, the device should have a dedicated **service interface**, e.g., via USB bus of the host MCU. Selection of a simple 8-bit MCU with USB (e.g., a PIC18 family member) will be required (*Note*: USB interface option is not mentioned Fig. 4 nor included in schematics). During manual device installation (see Fig. 5) this interface can be used by a PC or smartphone in combination with a dedicated software routine to submit a couple of suitable AT commands to the NB-IoT network module via MCU UART interface (see sample sequence used for "RasPi mockup and network tester" for reference). Goal is to fix an individual cellular network connection which is most appropriate for the individual device location. Usually, this process will be performed only one once during lifetime of deployed site, because locations of both device and selected cell tower will not change, i.e., configured connection will remain valid.

In addition, it might be helpful to implement a **signal quality LED indicator** to optimize device location during manual installation. For this purpose, after network attach the host MCU actual can use actual RSRP and RSRP values to visualize current quality of network connection of a NB-IoT device, e.g., translate them into a four-bar LED indicator or a multi-color LED to be driven by GPIOs of the MCU. This feature might be helpful during deployment of an IoT device when looking for the right location: the LED indicator will provide direction to cell tower resp. where to move the device for better signal strength.

Block Diagram/Schematics

Figure 8 is meant as a block diagram rather than schematics. It contains all major components and interconnections, but does not cover all details of the circuit, e.g., many passive components are just mentioned without a specific value, others are just left out, e.g., blocking capacitors.

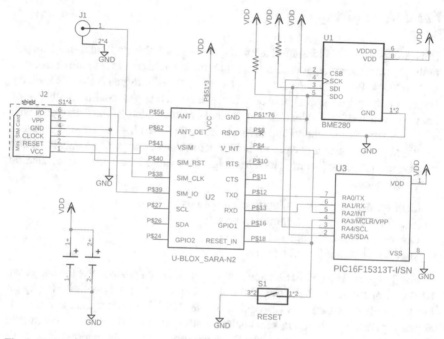

Fig. 8 Environmental sensor—block diagram/schematics

Board Assembly Layout

See Fig. 9.

Design Concept #2: Object Localizer

This design is representing the second group of predestinated NB-IoT target applications (see section "Object Tracking/Localization" of chapter "IoT Target Applications"). The idea is to use a dedicated NB-IoT device for remote determination of the actual geographic position of an object. Localization is based on GPS satellite positioning system. In fact, the IoT device will be used as a **protection instrument**—to be integrated into the target product by its manufacturer. I.e., described IoT device is an OEM product (OEM = Original Equipment Manufacturer) for B2B business. Protected object could be an expensive good or tool or equipment or shipment container, etc. In case of loss or theft or any other unintended or unexpected absence of the object, its owner (or an authorized person) can trigger our

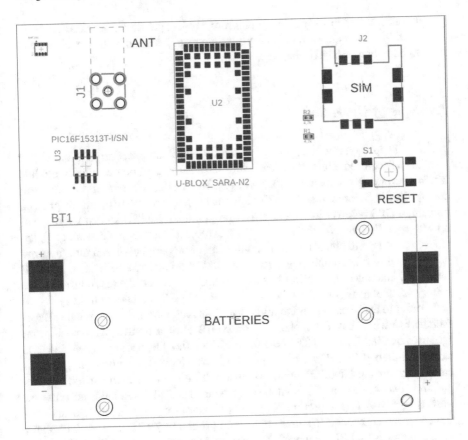

Fig. 9 Environmental sensor—PCB assembly layout

NB-IoT object localizer to return its actual position data, e.g., via SMS message. This information can be used for further follow-up or as an evidence for liability claims, for example.

In fact, for a reliable localization of an object the following requirements must be met:

1. Guaranteed data delivery,
2. Ubiquitous network coverage incl. good indoor penetration, and
3. Ultralong battery lifetime.

Worldwide 3GPP acceptance und deployment of regional cellular NB-IoT networks will ensure good coverage (see chapter "NB-IoT Network Deployment") and a reliable service. By nature, NB-IoT enhanced coverage capabilities will increase chance of connection in urban areas, even below ground level. Our object localizer should allow protection of any kind of target object, i.e., it has to work 100% autonomously, and it has to be powered by a non-rechargeable battery. The idea is to offer our object localizer as a protection add-on (OEM) product to manufacturers. Lowest power consumption is a key selling point for an object protection service for a committed period. This means that battery lifetime resp. installed battery capacity will have to be a focus design objective.

Principle of Operation

For easy integration into the protection target object a compact design of our object localizer is required. Here, the idea is to offer a fully autonomous add-on module which is working completely independently. In an effort to minimize component count, **Nordic Semiconductor nRF9160** has been selected. This chip integrates a modem, a GPS receiver as well as a user-programmable MCU so that we will end up with a single-chip NB-IoT solution for this project. An integrated Analog/Digital Converter is used to check battery condition, all other peripherals and interfaces are not used. Only few components need to be added: a battery, a SIM card, and LTE resp. GPS antennas. In an effort to save cost and to increase device robustness, the battery will be firmly mounted, i.e., soldered to the PCB (no battery holder).

In order to avoid cables and to allow seamless integration, an integrated surface-mounted antenna has been chosen: **NN03-310 chip antenna** component from Ignion to be used for both LTE/NB-IoT and GPS radio. Depending on selected LTE carrier frequency (see Fig. 1 of chapter "NB-IoT Network Deployment"), a dedicated filter for the NN03-310 chip antenna will be used for each carrier. For this purpose, two multi-port RF switches are being used to implement an **antenna matching network** (see section "Matching of Antenna"). These switches are controlled by the embedded IoT user application program by nRF9160 MAGPIO pins indicating the actual RF frequency. Proprietary AT commands need to configure this function before any modem activity occurs. Based on the given configuration, the modem applies the MAGPIO state corresponding to the RF frequency range automatically during runtime.

The nRF9160 has an ARM Cortex-M33 CPU core, 1 MB embedded flash memory and offers versatile support for development and debugging of an **IoT user application program** (see section "NB-IoT Cellular Network Modules" of chapter "Ingredients for NB-IoT Design Concepts" for a brief competitive overview of alternate options). For day-to-day operation, the IoT application will have to configure the modem in a way that it will connect to an appropriate local network wherever it will be located after each wake-up. Normally, you do not know where this will be. Thus, selection of an appropriate MNO partner is a key aspect to be considered during application design (see section "Roaming and MVNO" of chapter "NB-IoT Network Deployment"). The IoT application will have to execute an initial connection test after first boot during quality assurance (QA) step. When done, the device will enter an infinite loop until end of lifetime has been reached. The IoT application program will have to make sure that NB-IoT power saving features (esp. eDRX and PSM) are properly negotiated and applied. Focus is on periodic wake-up, TAU and upload of small data packages. Only limited reachability (e.g., daily) and infrequent short paging opportunities are required. User interactions should be limited to a minimum. In order to maximize battery life, position data is delivered on request only, i.e., the device works in "pull mode." Only **two SMS trigger command templates** for infrequent usage are needed (one for a "sign of life" and battery status, one as a GPS localization trigger). No physical human access or battery replacement is planned during product lifetime, i.e., device is built for "zero-touch" operation.

Of course, this configuration can be adjusted in order to tailor our object localizer for other application scenarios.

Functional Description

Figure 10 is illustrating the device operation from a software point of view. After comprehensive device configuration during production, each deployed device will be prepared for zero-touch operation during its complete product lifecycle and run in an infinite program loop.

Note: Numbers in brackets refer to program steps. Example: "(3)"

During production, our NB-IoT Object Localizer device will be prepared (1) for first wake-up after its deployment. In fact, it has to work anywhere, i.e., when triggered in the field it should connect to **any** NB-IoT (or LTE-M) network which is locally available. This initial configuration (2) should be verified (2), and—if successful—the network module can enter PSM sleep. In fact, PSM is the default operation mode for 24 h until the network module will wake up for a daily activity period (4). After wake up from PSM, the device will first have to connect to an available network cell (5), perform a TAU (tracking area update) and check for a pending SMS message which was addressed to the device during the last 24 h (6). In most cases, there will be no SMS, and the device can go back to sleep (4).

If a pending SMS is available for reception, it might just ask for the actual battery voltage as a "sign of life" from the device. For this purpose, nRF9160 offers a proprietary AT command which is returning measurement result with a resolution of 4 mV like this:

```
AT%XVBAT
%XVBAT: 3600
OK
```

Based on this value (here: "3600" which stands for 3000 mV), the host MCU will have to create a corresponding SMS message (7) and send it to a pre-configured recipient. If the received SMS was asking for the actual device position, the IoT application program will have to cold start the GPS engine (8) and send an SMS containing the GPS coordinates. When done, the device can return to PSM mode (9) and re-enter the infinite 24 h-loop (4) in PSM mode.

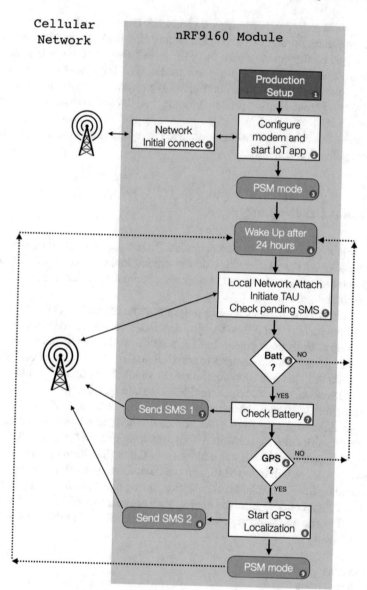

Fig. 10 Object localizer—application program flow

Note: SMS messaging is not supported by all NB-IoT networks. As an alternative, a low-power protocol should be used, e.g., MQTT (Message Queuing Telemetry Transport) or raw TCP socket-to-socket.

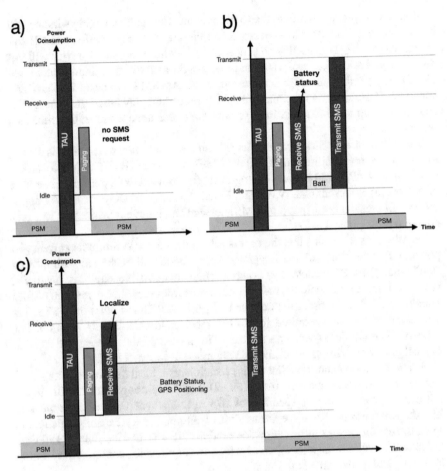

Fig. 11 Object localizer—timing/power overview

Power Consumption

Most of the time, our NB-IoT object localizer will be sleeping. Power consumption is determined by nRF9160 chip with a current supply of only 2.7 µA during PSM mode. Active daily device operation after waking up from default PSM is illustrated in Fig. 11. At the same time, the internal MCU will be activated so that the IoT application can resume and continue to control the IoT device incl. modem. Periodic TAU will typically take 2.7 s according to nRF9160 Product Specification [25]. At this time, the actual network coverage situation of the device and signal strength is unknown because the object might have been relocated since last TAU. Thus, worst case scenarios might apply (see section "Latency" of chapter "Cellular IoT Technology") for following network attachment procedure. Required time to retrieve relevant MIB/SIB network system information and subsequent execution of random access procedure might take 3 s maximum. Subsequent paging window will be checked for pending SMS messages containing user requests for our NB-IoT device. No message means that no action is required this time, so the device will just initiate a periodic TAU procedure and then return to PSM state immediately, see case (a). If a pending message is indicated, associated SMS will be downloaded and analyzed. Depending on outcome of PRACH procedure, download traffic might be applied with up to 2048 repetitions, so we have to calculate 2 s reception time as a worst case.

Two kinds of user SMS requests have been specified. In case (b), user is asking for actual battery condition as a "sign of life" from our NB-IoT object localizer. Consequently, nRF9160 internal analog-to-digital converter will perform a one-shot measurement of current battery voltage. Based on returned ADC data, the IoT application will create a corresponding SMS message and initiate transmission to a pre-configured recipient.

Finally, case (c) means that the user is asking the device to return its current geo position. For this purpose, the integrated GPS engine will be started with a "cold start" when the GPS receiver has no knowledge about visible satellites, positions, velocity, time, etc. As such, the receiver must start from scratch, e.g., systematically search for satellites in order to compute first position. "Time to first fix" (TTFF) is a measure of the time required for a GPS receiver to calculate its position. For nRF9160 internal GPS receiver a typical TTFF for a cold start is specified as 36 s. Typical power consumption during tracking mode is 44.9 mA [25].

Again, since we cannot predict device location resp. local network coverage, for our battery lifetime estimate we use nRF9160 peak modem current at 23 dBm transmit output power which is specified as 275 mA. For receive operation, we find 35 mA peak current. Now, we are ready to feed our battery capacity calculation spreadsheet (for further information see section "NB-IoT Device Battery Lifetime Calculator") with all these numbers. Further assumptions for configured operation of our object localizer are as follows:

- Device will wake up every 24 h, perform TAU, attach to a local NB-IoT network and will be available for paging.
- "Sign of life" request for actual battery status will be requested by user once per week (i.e., entered as periodic occasion of 52 per year in spreadsheet).
- GPS localization requests will be submitted 10 time during device lifetime (i.e., entered as 10 single occasions in spreadsheet).

Based on these assumptions and an IoT application program controlling modem and network functions accordingly, described object localizer device with an installed battery capacity of 3000 mAh will have an **estimated lifetime of 14.6 years**.

Of course, all device parameters incl. schedules, power consumption, and battery capacity can be adjusted and balanced differently in order to meet other requirements or use cases (Fig. 12).

Item	current (mA)	duration of single event (ms)	additional periodic occasions per day	total duration per scheduled day (ms)	total duration per day (hrs)	periodic occasions per year	total duration per year (hrs)	consumed capacity per hour (mAh)	periodic capacity consumption per year (mAh)	single occasion(s)	add capacity consumtion for single occasions (mAh)
periodic TAU, PRACH + paging	250	5700	0	5700	0,001583	365	0,578	0,01649	144,47917	0	0
receive SMS	35	2000	0	2000	0,000556	52	0,029	0,00012	1,01111	0	0
ADC + transmit battery status	275	10000	0	10000	0,002778	52	0,144	0,00453	39,72222	0	0
GPS + transmit location data	319,9	46000	0	46000	0,012778	0	0,000	0,00000	0,00000	10	0,1278
	0	0	0	0	0,000000	0	0,000	0,00000	0,00000	0	0
total ACTIVE time	879,9	0		63700	0,017694		0,751	0,021143	185,2		0,1278
nRF9160 in PSM mode	0,003										
total IDLE time (= 100% minus active time)	0,003			86336300	23,982306		8759,2	0,00270	23,6		
Subtotal periodic consumption									208,9		
TOTAL				86400000	24,000000		8760,00	0,02384			209,0
				86400s * 1000 = one day			8760hrs = one year		⇩		
battery capacity (mAh)	3000								battery lifetime in years:		14,4

Fig. 12 Object localizer—battery lifetime

Block Diagram/Schematics

Figure 13 is meant as a block diagram rather than schematics. It contains all major components and interconnections, but does not cover all details of the circuit, e.g., many passive components are just mentioned without a specific value, others are just left out, e.g., blocking capacitors.

L/C matching networks for LTE bands depend on selected NB-IoT carrier frequencies to be supported by design. Values are listed in [26]. In fact, 3GPP has specified around 25 NB-IoT bands so far. RF switches used here are supporting up to seven bands plus GPS only.

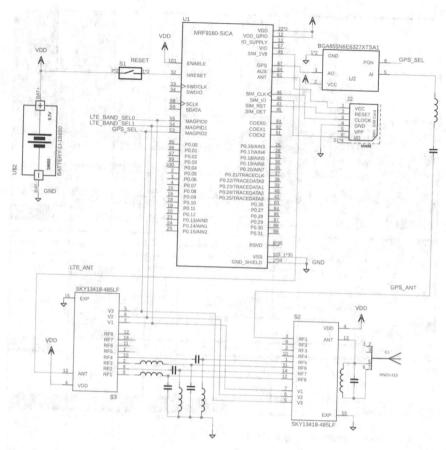

Fig. 13 Object localizer—block diagram/schematics

Board Assembly Layout

Figure 14 is showing the top view of a sample board containing all components of the block diagram (Fig. 13). In particular, it demonstrates compact board design, even if components are placed on the one side of the PCB only. For this component placement, board size is just 71.1 × 46.6 mm. With a double-sided assembly approach or a smaller battery, even smaller dimensions are possible. Here, for the form factor of the Lithium battery, so-called 18650 industry standard (referring to size of 65 × 18 mm) has been used which is offering a great variety of capacities of up to 10 Ah (!) available from many different manufacturers.

In general, for good RF performance a decent PCB layout is required. It is highly recommended to refer to related instructions and component values provided by Nordic Semiconductor and antenna manufacturer Ignion on their websites.

Fig. 14 Object localizer—
PCB assembly layout

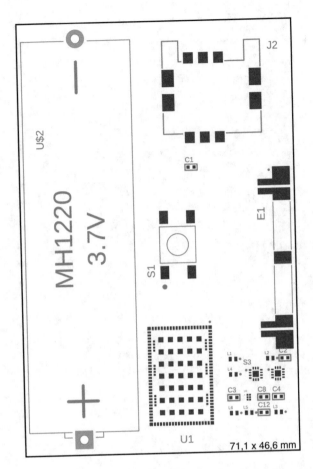

Glossary: Explanation of Frequently Used Cellular IoT Acronyms

3rd Generation Partnership Project (3GPP) Global standardization body for cellular mobile telecommunication protocols.

Cellular IoT (CIoT) IoT based on cellular network technology.

Common Criteria Evaluation Assurance Level (CC EAL) Certified security confidence level following a completed product evaluation according to common criteria rules.

Coverage Enhancement level 0–2 (CE level 0–2) NB-IoT method to improve network reach using repeated data transmissions.

Diffie–Hellmann (DH) Handshake protocol for secret key agreement to secure a channel.

evolved NodeB (eNodeB (eNB)) LTE base station.

Extended discontinuous reception (eDRX) Power saving feature (since 3GPP Rel 13).

General Packet Radio Service (GPRS) 2G/3G mobile data standard.

Global Positioning System, Global Navigation Satellite System (GPS, GNSS) Satellite-based radio navigation systems.

Hybrid Automatic Repeat Request (HARQ) LTE process combining data retransmission and error correction.

International Mobile Subscriber Identity (IMSI) Unique identifier for mobile user, assigned by MNO in SIM resp. eSIM.

IoT Cloud On-demand IT services to facilitate IoT devices incl. collection, analysis, processing of IoT data.

Low-Power Wide-Area Network (LPWAN) Category of wireless networks technologies incl. NB-IoT, etc.

Machine-to-machine (M2M) Direct communication between devices.

Maximum Coupling Loss (MCL) Maximum signal loss that a wireless system can tolerate and still be operational (for NB-IoT: −164 dBm).

Mobile Network Operator, Mobile Virtual Network Operator (MNO/ MVNO) Wireless communications services provider.

Narrowband IoT (NB-IoT) Extension of 3GPP LTE standard dedicated to typical IoT requirements.

Power Save Mode LTE feature (since 3GPP Rel 12).

Public Key Infrastructure (PKI) Entity for management of certificates for cryptographic keys used for IT security.

Public Land Mobile Network Unique ID for country and network operator.

Radio Access Technology (RAT) Connection method for a radio based communication network, e.g., LTE, Bluetooth, WiFi.

Radio Resource Control (RRC) Network protocol used between IoT device and Base Station.

Random Access (RA) Procedure initiated by user device to apply for data transfer.

Reference Signal Received Power (RSRP) Measured value of received power of the LTE reference signals.

Signal to Interference plus Noise Ratio (SINR) Quality indicator for a wireless connection.

Subscriber Identification Module (SIM) Smart card provided by network provider to authenticate user device.

Transport Layer Security (TLS) Cryptographic communication protocol (SSL successor).

References

1. Landström S, Bergström J, Westerberg E, Hammarwal D (2016) NB-IoT: a sustainable technology for connecting billions of devices. Ericsson Technol Rev
2. https://www.gsma.com/iot/mobile-iot/
3. LPWAN network whitepaper (2020) Telstra
4. 3GPP TR45820 (2015) Cellular system support for ultra-low complexity and low throughput Internet of Things (CIoT). 3GPP Technical Report
5. Martiradonna S, Piro G, Boggia G (2019) On the evaluation of the NB-IoT random access procedure in monitoring infrastructures. Politecnico di Bari, Italy
6. Cellular networks for Massive IoT (2020) Ericsson whitepaper
7. Narrowband Internet of Things (2016) Rohde&Schwarz Whitepaper
8. Muteba F, Djouani K, Olwal TO (2020) Challenges and solutions of spectrum allocation in NB-IoT technology. Tshwane University of Technology, Pretoria, South Africa
9. Harwahyu R, Cheng R-G, Liu D-H, Sari RF (2021) Fair configuration scheme for random access in NB-IoT with multiple coverage enhancement levels. IEEE Trans Mob Comput 20(4)
10. 3GPP TS 36.321. Evolved universal terrestrial radio access (E-UTRA); Medium Access Control (MAC) protocol specification. 3GPP Technical Specification
11. 3GPP TS 36.213. Evolved universal terrestrial radio access (E-UTRA); Physical layer procedures. 3GPP Technical Specification
12. Liberg O, Sundberg M, Wang E, Bergman J, Sachs J, Wikström G (2019) Cellular Internet of Things. Academic, London
13. GSMA (2019) NB-IoT deployment guide to basic feature set requirements
14. Power saving methods for LTE-M and NB-IoT devices (2019) Rohde&Schwarz Whitepaper
15. Feltrin L, Tsoukaneri G, Condoluci M, Buratti C, Mahmoodi T, Dohler M, Verdone R (2019) NarrowBand-IoT: a survey on downlink and uplink perspectives. IEEE Wirel Commun 26(1):78–86
16. Jorke P, Falkenberg R, Wietfeld C, Communication Networks Institute of TU Dortmund (2018) Power consumption analysis of NB-IoT and eMTC in challenging smart city environments
17. U-blox (2020) SARA-N2/SARA-N3 series - AT commands manual. UBX-16014887, Release R18
18. Telit (2021) ME310G1/ME910G1/ML865G1 AT Commands Reference Guide. 80617ST10991A Rev. 9
19. Common Criteria for Information Technology Security Evaluation (CC). https://www.commoncriteriaportal.org
20. Bar-El H. Known attacks against smartcards. Whitepaper. Discretix Technologies Ltd
21. Protection Profile for the Gateway of a Smart Metering System (Smart Meter Gateway PP) (2014) Version 1.3. www.bsi.de
22. Quectel (2019) BG96 - AT commands manual. Release V2.3

© The Author(s), under exclusive license to Springer Nature Switzerland AG 2022
K. Heins, *NB-IoT Use Cases and Devices*, https://doi.org/10.1007/978-3-030-84973-3

23. Quectel (2018) BG96 network searching scheme introduction. Release V1.2
24. Martinez B, Adelantado F, Bartoli A, Vilajosana X (2019) Exploring the performance boundaries of NB-IoT. IEEE, New York
25. Nordic Semiconductor (2020) nRF9160 – Product Specification. V2.0
26. Ignion (2021) APPLICATION NOTE TRIO mXTENDTM (NN03-310)

Printed in the United States
by Baker & Taylor Publisher Services